Palgrave Studies in Climate Resilient Societies

Series Editor
Robert C. Brears
Avonhead, Canterbury, New Zealand

The Palgrave Studies in Climate Resilient Societies series provides readers with an understanding of what the terms **resilience and climate resilient** societies mean; the best practices and lessons learnt from various governments, in both non-OECD and OECD countries, implementing climate resilience policies (in other words what is 'desirable' or 'undesirable' when building climate resilient societies); an understanding of what a resilient society potentially looks like; knowledge of when resilience building requires slow transitions or rapid transformations; and knowledge on how governments can create coherent, forward-looking and flexible policy innovations to build climate resilient societies that: support the conservation of ecosystems; promote the sustainable use of natural resources; encourage sustainable practices and management systems; develop resilient and inclusive communities; ensure economic growth; and protect health and livelihoods from climatic extremes.

More information about this series at
http://www.palgrave.com/gp/series/15853

Franziska Alesso-Bendisch

Community Nutrition Resilience in Greater Miami

Feeding Communities in the Face
of Climate Change

Franziska Alesso-Bendisch
Well Life Ventures
Miami Beach, FL, USA

ISSN 2523-8124 ISSN 2523-8132 (electronic)
Palgrave Studies in Climate Resilient Societies
ISBN 978-3-030-27450-4 ISBN 978-3-030-27451-1 (eBook)
https://doi.org/10.1007/978-3-030-27451-1

© The Editor(s) (if applicable) and The Author(s), under exclusive license to Springer Nature Switzerland AG 2020
This work is subject to copyright. All rights are solely and exclusively licensed by the Publisher, whether the whole or part of the material is concerned, specifically the rights of translation, reprinting, reuse of illustrations, recitation, broadcasting, reproduction on microfilms or in any other physical way, and transmission or information storage and retrieval, electronic adaptation, computer software, or by similar or dissimilar methodology now known or hereafter developed.
The use of general descriptive names, registered names, trademarks, service marks, etc. in this publication does not imply, even in the absence of a specific statement, that such names are exempt from the relevant protective laws and regulations and therefore free for general use.
The publisher, the authors and the editors are safe to assume that the advice and information in this book are believed to be true and accurate at the date of publication. Neither the publisher nor the authors or the editors give a warranty, expressed or implied, with respect to the material contained herein or for any errors or omissions that may have been made. The publisher remains neutral with regard to jurisdictional claims in published maps and institutional affiliations.

Cover illustration: © Melisa Hasan

This Palgrave Pivot imprint is published by the registered company Springer Nature Switzerland AG
The registered company address is: Gewerbestrasse 11, 6330 Cham, Switzerland

To Leonardo, Sebastian and Nicolas—may we find solutions for you be able to enjoy living in Miami for many more years to come

Acknowledgements

This book is based on research conducted in 2018 and 2019 in Greater Miami.

Most of all, I am grateful to more than 20 people I interviewed, listed in Appendix 1. These inspiring individuals from various sectors work tirelessly to keep Miami liveable and enjoyable for future generations, despite numerous challenges posed by the political environment, economic constraints, and the changing climate (to name just a few). I can't thank them enough for their openness and dedication.

In Miami-Dade County, I'd like to thank Jim Murley, Jane Gilbert, Susanne Torriente, Commissioner Daniella Levine-Cava, Mayor Phil Stoddard, Charles Lapradd, Dr. Jaap Donath, Irela Bagué, Zelalem Adefris, Thi Squire, James Jiler, Juan Casimiro, Kate Stein, Art Friedrich, and Rebecca Fishman-Lipsey.

In Broward County, I'm grateful to Jennifer Jurado and Jill Horwitz.

In Palm Beach County, I'd like to thank Megan Houston.

Working across (and beyond) county borders, I thank Paco Velez, Gretchen Schmidt, Jeannie Necessary, Yoca Arditi-Rocha, Dr. Janisse Schoepp, Elisa Juarez, Jose Caceres, and Chris Castro.

A special thank you to the people who reviewed early drafts of the manuscript and provided their feedback. I am grateful to you, especially to Kate Stein, Dr. Rachele Hendricks-Sturrup, and Jeannie Necessary.

Finally, I want to acknowledge with gratitude the support of my family here and abroad and particularly of my husband, Alvaro, without whose encouragement and help I would likely not have embarked on this extensive yet exciting project. Thank you to my children, Leo, Sebas, and Nico, who only have enforced my ambition to do my part for the health of our communities and our planet, so they will continue to be able to enjoy Miami as much as we do now.

Contents

1 **Prologue: Community Nutrition Resilience—What and Why**   1
   1.1   *Resilience of Social-Ecological Systems*   4
     1.1.1   Resilience—A New Term for Sustainability?   6
     1.1.2   Resilience, Adaptability, and Transformability   7
     1.1.3   Urban Resilience   8
     1.1.4   Community Resilience   12
     1.1.5   Climate Adaptation   14
   1.2   *Food Security and Climate Change*   17
   1.3   *Food Systems: The Vehicle for Community Nutrition Resilience*   21
   1.4   *Community Nutrition Resilience: Moving from "Survive" to "Thrive"*   29
   References   29

2 **Resilience Challenges to Community Nutrition Security in Greater Miami**   37
   2.1   *Historical Dynamics*   39
     2.1.1   Political Support   39
     2.1.2   Immigration   41
     2.1.3   Community Development   43
     2.1.4   Food System   45
     2.1.5   Climatic Events   46

2.2 Climatic Stresses and Shocks 47
    2.2.1 Threats to Economy 47
    2.2.2 Human Impacts 50
2.3 Status Quo 51
    2.3.1 Community Resilience 51
    2.3.2 Food Security 55
    2.3.3 Health Implications 59
2.4 Threats to Community Nutrition Resilience in Miami 60
    2.4.1 Lack of Focus, Policy, and Planning 60
    2.4.2 Threats Related to the Food System 62
    2.4.3 Health Literacy and Education 67
    2.4.4 Adaptive Capacity 68
References 69

3 Taking (Community Nutrition) Resilience Action 75
3.1 Political Action 77
    3.1.1 Federal 77
    3.1.2 State 79
    3.1.3 Greater Miami 79
3.2 Regional Resilience Action 80
    3.2.1 Southeast Florida Regional Climate Change Compact 80
    3.2.2 100 Resilient Cities (100RC) 82
    3.2.3 Resilient305 83
3.3 Miami-Dade County 84
    3.3.1 City of Miami 86
    3.3.2 City of Miami Beach 88
    3.3.3 Miami Gardens 89
    3.3.4 South Miami 90
3.4 Broward County 90
    3.4.1 Fort Lauderdale 95
    3.4.2 Other Major Cities in Broward County 96
3.5 Palm Beach County 97
    3.5.1 West Palm Beach 97
    3.5.2 Boca Raton 98
3.6 Community Engagement and Other Key Initiatives 98
    3.6.1 Key Nonprofit Organizations 99
    3.6.2 Community Foundations 102

3.6.3	Academia	103
3.6.4	Private Sector	103
3.6.5	Media	105
3.6.6	Civil Sector	105
References		106

4 Designing Nutrition Resilient Communities: Learnings from Other Cities   111

4.1	Boston	112
4.2	New York City	115
4.3	Baltimore	116
4.4	Orlando	118
4.5	Los Angeles	120
4.6	Other Cities	121
4.7	Food Waste and Recovery	123
References		124

5 Conclusions—Making Greater Miami's Communities Nutrition Resilient   127

5.1	Recommendations Related to Planning and Policy	128
	5.1.1 Form a Central Entity Responsible for Community Nutrition Resilience	128
	5.1.2 Conduct Assessment of the Status Quo of Community Nutrition Resilience	130
	5.1.3 Map the Food System, Identify Health Disparities, and Support the Business Case	131
	5.1.4 Conduct Food System Resilience Assessments	134
	5.1.5 Develop a Vision, Strategy, and Aims for Community Nutrition Resilience	135
	5.1.6 Incorporate Food Systems and Community Nutrition Resilience into Resilience Planning	136
	5.1.7 Co-create Neighborhood Nutrition Resilience Plans	137
	5.1.8 Develop Government Policies and Practices that Help the Food System Quickly Return to Normal Operations	138
5.2	Recommendations to Strengthen Food System Resilience	140
	5.2.1 Food Production	141

		5.2.2	Food Processing and Distribution	145

      5.2.2   *Food Processing and Distribution*   145
      5.2.3   *Food Access*   146
      5.2.4   *Health Literacy and Community Education*   149
      5.2.5   *Cultural Relevance*   150
      5.2.6   *Food Waste and Recovery*   150
  5.3 *The Way Forward: From Fun in the Sun to Resilience Excellence*   151
  5.4 *Areas for Further Research*   153
  5.5 *Conclusions*   154
  References   156

Appendix A: List of Interview Partners (Alphabetic Order)   159

Appendix B: RCAP Actions by Municipality (Southeastfloridaclimatecompact.org 2018d)   161

Bibliography   189

Index   211

Abbreviations and Acronyms

&	And
100RC	100 Resilient Cities
A/C	Air Conditioning
A/S	Aksjeselskap (Norwegian term for a stock-based company)
AI	Artificial Intelligence
AIA	The Miami Chapter of American Institute of Architects
ALICE	Asset Limited, Income Constrained, Employed
AMA	American Medical Association
BFPI	Baltimore Food Policy Initiative
BMI	Body Mass Index
CCE	(Office of) Civic and Community Engagement
CDC	Centers for Disease Control and Prevention
CGIAR	Consortium of International Agricultural Research Centers
CO_2	Carbon dioxide
CRO	Chief Resilience Officer
CSA	Community-Supported Agriculture
CSN	Corporate Sustainability Network
D.C.	District of Columbia/Federal District
D-Snap	Disaster Supplemental Nutrition Assistance Program
e.g.	Exempli gratia (lat.)/for example
EMD	Emergency Management Department
EOC	Emergency Operations Center
EPA	United States Environmental Protection Agency
ERS	Economic Research Services
et al.	Et alia (lat.)/and others
FAB	Fresh Access Bucks program

FANM	Family Action Network Movement
FAO	Food and Agriculture Organization of the United Nations
FDA	Food and Drug Administration (of the USA)
FEMA	Federal Emergency Management Agency
FIU	Florida International University
FLIC	Florida Immigrant Coalition
FMI	Food Marketing Institute
FPL	Florida Power & Light
FRAC	Food Research and Action Center
GHG	Greenhouse Gases
GIS	Geographic Information System
GM&B	Greater Miami and the Beaches
Gov.	Governor
HOLC	Home Owners Loan Corporation
HUD	United States Department of Housing and Urban Development
i.e.	Id est (lat.)/in other words
ICIC	Initiative for a Competitive Inner City
IFAS	Institute of Food and Agricultural Sciences
IISD	International Institute for Sustainable Development
IPCC	Intergovernmental Panel on Climate Change
IWRA	International Water Resources Association
LED	Light-Emitting Diode
M	Million
MCA	Miami Climate Alliance
MDC	Miami-Dade County
MHA	Miami Housing Authority
MIT	Massachusetts Institute of Technology
MWC	Miami Workers Center
NAFTA	North American Free Trade Agreement
NIFA	National Institute of Food and Agriculture
OFI	Office of Food Initiatives
PBC	Palm Beach County
Ph.D.	Doctor of Philosophy
RCAP	Regional Climate Action Plan
SAP	Sustainable Action Plan
SES	Socio-Ecological System
SFWMD	South Florida Water Management District
SLR	Sea-Level Rise
SNAP	Supplemental Nutrition Assistance Program
SNAP-Ed	Supplemental Nutrition Assistance Program Education
SR	State Road
SSB	Sustainability Stewards of Broward

U.S.	United States (of America)
UF	University of Florida
UGW	Urban Green Works
UK	United Kingdom
ULI	Urban Land Institute
UM	University of Miami
UN	United Nations
US$	US Dollar
USA	United States of America
USC	University of South Carolina
USDA	United States Department of Agriculture
WHO	World Health Organization
WIC	Special Supplemental Nutrition Assistance Program for Women, Infants and Children

List of Figures

Fig. 1.1	Dimensions of community food security (Adapted from Frayne et al. 2010)	18
Fig. 1.2	The food system (Adapted from Zeuli and Nijhuis 2017)	24
Fig. 2.1	How sea level rise could affect South Florida in 2100 (Climate Central 2018)	47
Fig. 2.2	Miami-Dade county 2014. Residents living below poverty (data analysis and mapping by FIU Metropolitan Center 2016)	52
Fig. 2.3	Low supermarket sales and low income (The Food Trust 2012)	58
Fig. 5.1	Areas with greatest need (The Food Trust 2012)	133

LIST OF TABLES

Table 1.1   Potential impacts of climate change on food systems
            and food security   22
Table 2.1   Selected food security parameters for Miami-Dade,
            Broward, and Palm Beach Counties   57
Table 2.2   Selected health parameters for Miami-Dade, Broward,
            and Palm Beach Counties   60
Table 5.1   Challenges to Greater Miami's food system   134

CHAPTER 1

Prologue: Community Nutrition Resilience—What and Why

Abstract This chapter conceptualizes community nutrition resilience. It draws on a literature review related to resilience, delineating the term from the related terms "sustainability," "transformability," and "adaptability." It examines resilience in an urban setting and then among communities. Subsequently, it looks at food security and identifies nutrition as a critical component affecting the health, well-being, and opportunities of communities. It pays particular attention to how a changing climate affects food security. It then examines food systems and its constituent parts as the vehicle for delivering community nutrition resilience. Ultimately, it offers a definition of community nutrition resilience.

Keywords Resilience · Urban resilience · Community resilience · Food security · Food system · Community nutrition resilience

Nutritious food is the basis for human growth. Access to nutritious and safe food, most commonly defined as food security, determines in its worst case whether to live or die. Still, in a more moderate scenario, food security determines whether one can fully biologically develop, learn, and create, and benefit from economic opportunity.

When discussing food security, reference is usually made to developing countries and a lack of availability of or access to food. And in fact, around 827 million people worldwide lacked access to adequate foods

© The Author(s) 2020
F. Alesso-Bendisch, *Community Nutrition Resilience in Greater Miami*, Palgrave Studies in Climate Resilient Societies,
https://doi.org/10.1007/978-3-030-27451-1_1

in 2017 (FAO.org 2018), and about one-third of the global population suffer from deficiencies in essential vitamins and minerals (WHO 2019). However, food insecurity does not only affect developing countries. In the United States of America (USA), about 48 million people (or 15% of the population) live in food insecure households.

While under-nutrition is a serious challenge, millions of more people over-consume foods and beverages that are high in nutrients of concern, such as sugar and fat. In 2016, more than 1.9 billion adults, 18 years and older, were overweight. Of these, over 650 million were obese (WHO.int 2018a). In the USA, over 72 million people (or one in five) are obese, with obesity disproportionately affecting minority and low-income individuals (Kim and von dem Knesebeck 2018). As will be argued in this book, this is because these individuals tend to lack the money, the health literacy and skills, and sometimes even the physical access to the right stores, to be able to choose and prepare healthy foods such as fruits and vegetables.

Poor nutrition affects societies' ability to thrive. Obesity is one of the leading causes of life-threatening diseases including cardiovascular diseases (mainly heart disease and stroke), musculoskeletal disorders and some cancers (including endometrial, breast, ovarian, prostate, liver, gallbladder, kidney, and colon) (WHO.int 2018a). In addition, poor nutrition affects mental health (Harvard Health Publishing 2015) and has been linked to developmental deficits in children (Korenman et al. 1995). Economically, health costs both for the public and private sector have been soaring worldwide, increasing at a faster pace than the economy of most countries (WHO.int 2018b).

Overweight and obesity, as well as their related noncommunicable diseases, are largely preventable through living a healthy lifestyle that includes regular movement and a diet largely consisting of health-promoting foods such as fruits and vegetables, whole grains, and high-quality protein. Hence for prevention to happen, one must ensure that people are not only food secure in a sense of calorie secure, but nutrition secure as well. Nutrition security includes having physical and economic access to food that is healthy and culturally relevant and being empowered to make healthy food choices as it is often a lack of education that prevents people from choosing nutritious food over nutrient bankrupt alternatives.

Insight about why and how people become nutrition insecure suggests ways of preventing the physical, mental, developmental, and economic impacts of poor nutrition from happening. If interventions are designed in ways that increase nutrition resilience by enhancing people's ability both to make healthy choices every day and to deal with disturbances, then the need for interventions will diminish. Expanding insights on both is the primary goal of this book.

In a world of growing complexity and uncertainty, the security of food supplies and nutritional value of food is threatened by many factors. On the one hand, multiple processes of global change are happening over time such as urbanization, population aging, and the proliferation of Digital and Artificial Intelligence (AI) can impact the food system. On the other hand, there are unexpected shocks such as natural disasters and financial and political crises, threatening the security of our food supply.

Climate change, or the warming climate, is one of the biggest contributors to both chronic stresses and acute shocks threatening the resilience of food supply and quality. The warming climate affects the food system, from production to consumption and beyond. Further, it exacerbates the negative effects of (food-related) inequalities such as food deserts, to take just one measure of the lack of access to healthy and nutritious food.

For several decades now, resilience has found its way into urban planning and thus into the conversation about the ability of cities to cope with chronic stresses as well as acute shocks, particularly those related to climate change. However, the resilience of food systems—as a subsystem of cities and the vehicle to ensure nutrition security—is often not addressed in urban planning and programming. And that is despite research, for example, by the Initiative for a Competitive Inner City (ICIC), which argues that city leaders should include food systems as part of their resilience planning in order to avoid extended food supply disruptions in the aftermath of any type of upheaval (Whalen and Zeuli 2017). In addition, there's the systemic challenge of food deserts that needs to be addressed.

Although the investigation of cities and the activities of players in a complex urban system seem useful, looking at resilience on a community level makes sense for various reasons. First, food systems are very particular to neighborhoods and communities. For example, the decision of which type of store to open in a neighborhood, or what products to offer, is often determined by socio-demographic factors such as the income of that particular neighborhood. Second, interventions are oftentimes driven by community centers and on the community level.

Planning for community nutrition resilience, either as part of climate resilience initiatives or as within urban planning, can be simplified by operationalizing concepts of community resilience and nutrition resilience. Hence, this book will focus on nutrition resilience of Greater Miami's communities, both in the face of chronic stresses and acute shocks, particularly related to climate change.

This chapter will conceptualize the key term of community nutrition resilience. To do that, it will discuss the key concept of resilience—including its distinction from sustainability, adaptability, and transformability—and transfer it to cities and communities. Further, it will delineate nutrition security to ultimately arrive at a definition of community nutrition resilience. It will further look at the resilience of food systems, as a sub-system of community resilience. It will do so by reviewing the relevant literature from a number of related fields to result in a definition and conceptualization of community nutrition resilience that can aid the discussion for academia and practitioners alike.

1.1 Resilience of Social-Ecological Systems

For social-ecological systems, looking at resilience is key to establish their ability to thrive despite chronic stresses and acute shocks. However, many have argued that resilience is in danger of becoming just another buzzword (e.g., Davoudi 2012; Porter and Davoudi 2012; Longstaff et al. 2010) or that it has become an umbrella term, which loosely expresses some of the conceptual underpinnings of the adaptation approach taken (Fünfgeld and McEvoy 2012). For that not to happen, the term needs to be clearly defined and delineated from other terminologies such as sustainability, adaptability, and transformability.

Historically, the concept has evolved significantly. Resilience stems from the Latin word *resilire*, meaning to spring back, and has its origins in physics, where it is used to describe the stability of materials and their resilience to external shocks (Davoudi 2012). Since the 1960s, resilience has been used in the field of ecology, where multiple meanings of the concept have since emerged, with each being rooted in different world views and scientific traditions (Davoudi 2012). Holling (1973) first made the distinction between so-called engineering resilience and ecological resilience. The distinction of both of these views, with a third view of evolutionary resilience introduced below, is critical to the understanding of resilience in the context of a social system, such as the region Greater Miami, the subject of this book.

Engineering resilience has been defined as the ability to return to an equilibrium or "normal", steady state after a disturbance (Holling 1973). An example is an antenna that bends in heavy winds but returns to its normal position once the storm has passed. The measure of resilience in this view is the resistance to disturbance and the speed by which the system returns to equilibrium. This perspective is built on the assumptions of positivist social science with the existence of blueprints, certainty, and forecasting (Porter and Davoudi 2012).

Ecological resilience, on the other hand, has been associated not only with how long it takes to bounce back after a disturbance, but also how much disturbance someone, or something, can absorb before it changes its state (Holling 1996). Thus, it is understood not only as "the ability to persist, but also the ability to adapt" (Adger 2003: 1), bringing in a notion of learning and development, or the ability to bounce back versus to bounce forward (Shaw 2012). Hence, although both perspectives are based on the belief in the existence of some sort of equilibrium, there is one main difference between the two. Ecological resilience rejects the existence of a single, stable equilibrium and, instead, the existence of multiple equilibria and the possibility of systems to adapt alternative state of beings after a shock (Davoudi 2012).

As an alternative third perspective, contributors such as Folke et al. (2010) and Davoudi (2012) have since introduced evolutionary resilience. This perspective challenges the belief in an equilibrium and advocates that the very nature of systems may change over time with or without an external disturbance (Scheffer 2009). This view recognizes that the seemingly stable state that we can observe around us can suddenly change and become something radically new, with characteristics that are profoundly different from those of the original (Kinzig et al. 2006).

The evolutionary perspective broadens the engineering and ecological view of resilience to incorporate the dynamic interplay of persistence, adaptability, and transformability across multiple scales and time frames (Holling and Gunderson 2002; Walker et al. 2004; Folke et al. 2010). Resilience is no longer conceived as a return to normality, but rather as the ability of complex socio-ecological systems (SESs) to change, adapt, and crucially transform in response to stresses and strains (Carpenter et al. 2005). Systems are defined as "complex, non-linear, and self-organizing, permeated by uncertainty and discontinuities" (Berkes and Folke 1998: 12). SES, which includes cities and communities, is an interaction of several sub-systems, including human and natural systems. The evolutionary perspective on resilience has brought the role of institutions, leadership, social capital, and social learning into the scope of resilience (Olsson et al. 2006). While the origins of the term resilience imply strength and resistance, its application in SESs and urban sustainability understands that it requires flexibility, learning, and change (Tyler and Moench 2012).

When looking at community nutrition resilience in Greater Miami, this book will adopt the evolutionary perspective on resilience and apply it to look at the food system as one of its critical sub-systems. What will be looked at is how gradual, yet cumulative changes such as immigration, aging, or wealth distribution change the state of a region over time and affect its ability to bounce back and bounce forward.

In terms of a definition, the one used by scholars at the multi-disciplinary Resilience Alliance seems most relevant because it is most applicable to the investigation of the systems examined in this book (Resalliance. org 2018: 1):

> Resilience is the capacity of a social-ecological system to absorb or withstand perturbations and other stressors such that the system remains within the same regime, essentially maintaining its structure and functions. It describes the degree to which the system is capable of self-organization, learning and adaptation.

Consequently, resilience of SESs has three defining characteristics: (1) the amount of change the system can undergo and still retain essentially the same structure, function, identity, and feedback on function and structure; (2) the degree to which the system is capable of self-organization; and (3) the degree to which the system expresses capacity for learning and adaptation (Resilience Alliance 2010).

1.1.1 Resilience—A New Term for Sustainability?

Some argue that the term resilience has been replacing sustainability in everyday discourses (e.g., Davoudi 2012), and even if you don't want to go that far, it has certainly found its place in the conversation about climate change adaptation. Perhaps this is due to its less politically charged use and malleability (Davoudi 2012). To fully understand its relationship with sustainability, it is necessary to clearly define it and then delineate the two.

Sustainability, or sustainable development, has been most commonly defined as the ability to meet the needs of the present generation without compromising the ability of future generations to meet their needs (Brundtland 1987). Like resilience, sustainability recognizes the need for precautionary action on resource use and on emerging risk: the avoidance of vulnerability and the promotion of ecological integrity into the future (Adger 2003). However, in an urban planning and policy context, sustainability has been found to focus on climate change mitigation, mainly dealing with limiting greenhouse gas (GHG) emissions. Resilience, on the other hand, deals with the response to climate change and its consequence such as sea level rise or severe weather events.

Both resilience and sustainability deal with the same questions of justice, equity, fairness, and legitimacy (i.e., whose resilience is prioritized). Justice in resilience needs to account for the outcomes of resource allocations and policy decisions. Equity similarly describes fairness, in a sense that each individual or group is being met where they are. It's about the fairness of how societies can build the capacity to adapt when vulnerable groups are marginalized and excluded from decisions. It's also about fairness in procedures and institutions, recognition of difference, and participation in decision making (Adger 2003).

1.1.2 Resilience, Adaptability, and Transformability

According to Walker et al. (2004), the stability of all SESs emerges from three complementary attributes: resilience, adaptability, and transformability. In fact, adaptability is part of resilience (Folke et al. 2010), whereas transformability is not.

Adaptability represents the ability to adjust responses to changes in the external and internal environment. It is the capacity of actors inside the system to influence resilience or to manage it (Walker et al. 2004). These are usually human actors and their intent and capability determines the ability of the system to avoid crossing a threshold into an undesirable state. If chance favors the prepared mind, then preparing SESs for inevitable, yet uncertain change, must involve building adaptive capacity. Building the capacity to adapt is complementary to building resilience, as it increases options for re-organization following a disturbance (Quinlan 2003). However, much of cities' work on resilience has been called adaptation in practice.

Transformability, on the other hand, is the capacity to create a fundamentally new system when disturbances, or evolutionary change, have made the current system unsustainable (Walker et al. 2004).

Knowing if, when and how to initiate transformative change before it is too late to escape a seriously undesirable state, is key to SES transformability (Walker et al. 2004). Thereby, the concept of tipping points is a useful viewpoint. The question here is how much disturbance can a system take before it switches, completely or in parts, to an alternative state and discussing tipping points can hereby be useful to identify critical thresholds pointing to the need for more drastic adaptation measures (Fünfgeld and McEvoy 2012).

Human actors within a system need to focus on creating or maintaining the resilience of the current system and at the same time build capacity for transformability, should it be needed (Walker et al. 2004). Hence, whereas resilience and adaptability are related to the current boundaries of a system, transformability refers to fundamentally altering the nature of a system (Walker et al. 2004).

1.1.3 Urban Resilience

Urban resilience has been widely discussed in literature as the discussion of the existing body of research below demonstrates. However, there is a gap between the advocacy of urban resilience in the scientific literature and its take-up as a policy discourse on the one hand and the demonstrated capacity to govern for resilience in practice on the other (Wilkinson 2012).

Following the previously introduced engineering perspective on resilience, urban resilience has often been defined as "the capacity of a city to rebound from destruction" (Vale and Campanella 2005), be it related to its population, economy, or built form (Davoudi 2012). Similarly, a focus has been put on resilience as a buffer capacity for preserving what we have and recovering to where we were (Folke et al. 2010). This mirrors the prevailing emphasis in urban planning on the return to the "normal" state, e.g., pre-disaster. This is very much why much of the resilience-building literature, as well as practices, are dominated by post-disaster emergency planning, instead of taking a broader perspective that enables a stronger focus on progress and improvement.

In the USA, 100 Resilient Cities (100RC), an international organization funded by the Rockefeller Foundation and dedicated to helping cities become more resilient, has been instrumental in fostering the discussion about, and actions on urban resilience. They expand the traditional perspective with its definition of urban resilience as "*the capacity of individuals, communities, institutions, businesses, and systems within a city to survive, adapt, and grow no matter what kinds of chronic stresses and acute shocks they experience*" (100ResilientCities.org 2018).

For urban planning, this means that instead of viewing resilience as bouncing back to an original or "normal" state following an external disturbance, it should be seen as the ability to bounce forward, reacting to shocks by changing to a new, improved state that is more sustainable and resilient in the current environment (Shaw 2012). With this view, "blue-print" planning (Wilkinson et al. 2010: 31), while important, is no substitute for *"great leadership and a culture of teamwork and trust which can respond effectively to the unexpected"* (Seville 2009: 11). This acknowledges the importance of the ability to improvise or to use imagination (Shaw 2012) and encourages practitioners to consider innovation and change to aid recovery from disturbances and stresses that may or may not be predictable (Tyler and Moench 2012). Instead of focusing on seeking desirable states, planning should focus on resilience analysis, with a simultaneous focus on adaptive resource management and adaptive governance. Adaptive governance thereby is the process of creating adaptability and transformability in SESs (Walker et al. 2004).

In a resilient system, disturbance has the potential to create opportunity for doing new things, for innovation and for development. In a vulnerable system, on the other hand, even small disturbances may cause dramatic social consequences (Folke 2006). In a resilient region, for example, markets and local political structures continually adapt to changing environmental conditions and only when these processes fail is the system forced to alter the big structures (Swanstrom 2008).

Instead of focusing too much on reducing the uncertainty related to choosing a prediction to rely on and focusing on discreet measures to adapt to specific perceived future climate risks, it has been argued that it is more effective for cities to focus on building resilience (e.g., Tyler and Moench 2012). In fact, recent focus on resilience marks this shift from resistance strategies focused on the prediction of risk and the mitigation of vulnerability, to more inclusive strategies that integrate both resistance and resilience in the face of disaster (Longstaff et al. 2010). Still, resilience is often confused with resistance, strategies of which include physical countermeasures such as implementing pumps to stop flooding, or raising streets. Resilience strategies, on the other hand, assume that resistance may not always be possible and thus include the provision of, or access to, alternative resources and services if the resistance strategy fails. In that sense, resilience subsumes resistance (Longstaff et al. 2010).

For a city, resilience can be understood as a process that takes place in three sectors: private, public, and civic (or nonprofit) (Swanstrom 2008). Particularly, the role of governments, including federal, state, and local authorities, needs to be considered as well as that of policy makers. Sovereign authorities can play a role in both nurturing and undermining resilience (Swanstrom 2008). In fact, although the public sector represents the broadest scale and the slowest moving sector (Swanstrom 2008), government is critical to build the infrastructure of resilient systems like cities.

Looking at the underlying parts of urban resilience, Tyler and Moench (2012) identify three generalizable elements: systems, agents, and institutions. Based on these, they propose a conceptual framework that will be discussed below. This framework can inform the conceptualization of community nutrition resilience.

Systems are defined as the "*underlying support systems that enable networks of provisioning an exchange for urban populations*" (Tyler and Moench 2012: 313). They include both infrastructure and ecosystem services, such as regional food production, runoff management, or flood control. Food systems, as the vehicles of delivering nutrition resilience, will be discussed further below in this chapter. Systems can be further distinguished as core or critical systems, essential to human well-being, including water and food supply, energy, transport, shelter, and communications. Often, the failure of one of these systems can trigger the failure of others as they are interlinked (Tyler and Moench 2012).

According to Tyler and Moench (2012), resilient systems often demonstrate the following characteristics: flexibility and diversity (diversity defined in terms of place and function); redundancy and modularity (including spare capacity and a variety of options for service delivery); and safe failure (defined as both the ability to absorb shocks in ways that avoid catastrophic failure and the unlikeliness that the failure of one system leads to the failure of other systems) (Little 2002).

Agents are actors within a system that act with volition and intent (Tyler and Moench 2012). They determine the SES' adaptive capacity, as previously discussed. These agents can be individuals (e.g., farmers, consumers), households, and private- or public-sector organizations and have distinct interests and motivations, which can change based on experience, learning, and incentivization. Therefore, agents' resilience is determined by certain characteristics, including their responsiveness to likely disruptions and their capacity to act upon it; their resourcefulness, including their capacity to mobilize resources and access assets; and finally, their ability to learn and grow (Tyler and Moench 2012).

Finally, the concept of institutions refers to the social rules or conventions that structure human behavior and exchange in social and economic interactions (Hodgson 2006). They condition the way systems and agents interact to respond to climate-induced stress (Tyler and Moench 2012). Examples of institutions in the context of resilience are public information about hazards, inclusion in decision making, and ability to organize in preparation to or after external shocks, or pricing structures for urban services that influence access to infrastructure or food systems. Tyler and Moench (2012) offer the following characteristics that influence resilience related to institutions: rights and entitlements linked to system access, as they determine resilience particularly for marginalized groups; decision-making processes and particularly ensuring that the communities most affected have legitimate input in decision making (Huntjens et al. 2012); information flows and access to timely, credible, and meaningful information; and application of new knowledge (institutions that facilitate learning and adoption of new knowledge).

Following the above conceptualization, cities are vulnerable (as opposed to resilient) when fragile, inflexible systems and/or marginalized or low-capacity agents are exposed to increased hazards such as climate change, and their ability to respond is limited by constraining institutions (Tyler and Moench 2012). Resilience, on the other hand, is high, where robust and flexible systems can be accessed by high-capacity agents and where that access is enabled by supporting institutions.

One of the challenges in translating resilience to any society relates to power and politics and the conflict over questions such as, what is a desired outcome and resilience for whom (Davoudi 2012)? Resilience building in human systems is inherently conflictual or political (Swanstrom 2008) as some people gain while others lose. Hence, any discussion of resilience needs to consider issues of justice and fairness in terms of both the procedures for decision making and the distribution of burdens and benefits (Davoudi 2012). Further, issues of culture reference point and power relations need to be considered (Tyler and Moench 2012) as they inevitably constrain decision-making processes and options.

1.1.4 Community Resilience

This book argues that it is necessary to consider resilience on a community level, as opposed to the level of cities. A community is a group of people who share a common physical environment, resources and services, as well as risks and threats. It is also a collective body that has boundaries (often geographic), internal and external feedback, and "*a shared fate*" (Norris et al. 2008, 128). Looking at resilience on this level makes sense because, first, communities do exist outside of urban boundaries, also in rural areas. Secondly, community is an appropriate level for building basic resilience, particularly when looking at resilience against disasters. Most disasters are local and affect communities differently, such as hurricane Katrina and its devastating effect on New Orleans in 2005 or an earthquake. Furthermore, communities are unique and have their own local needs, experiences, systems, endowment of resources, and policies and ideas about prevention of, protection against, response to, and recovery from different types of disasters (Longstaff et al. 2010). They have access to resources and the ability to mobilize them that single individuals have not (Longstaff et al. 2010). A community-level focus on resilience results in local participation, ownership, and flexibility in building resilience.

Systems and their sub-systems are oftentimes particular to communities. In addition, each community has a "footprint", which can be defined as the region from which it pulls its resources, such as workers, water, or agricultural products (Longstaff et al. 2010). As described by Swanstrom (2008) for geographical regions in general, communities have porous boundaries with a constant flow of goods and people in and out. And they are subject to constantly changing influence factors that can either hinder or aid advancement. For this reason, communities depend heavily on their ability to adapt and reinvent themselves in order not to fall behind. Typically, it is the resources available to a community that facilitate these adaptations and allow communities to advance with the changes around them.

However, even with unlimited resources, it is highly unlikely that a community can prevent or protect itself from all the possible dangers it may face. In the United States, for example, complex distribution systems are now the primary mechanism for supplying populations with food and water. Communities build their everyday activities, as well as their means to adapt to changes, around complex systems centered around electricity, computerized systems, and communication networks (Longstaff et al. 2010), over which they have little or no control. Efficiencies inherent with modern technology and these complex systems of modern life reduce resilience through a loss of redundancy and diversity (Longstaff et al. 2010). An example is that most households nowadays have no home phone anymore, but rely on their mobile phones. When mobile telecommunication networks break down, there is no backup to establish communication.

As previously argued, much focus on resilience both in theory and praxis has centered around the response to acute shocks, including post-disaster recovery and risk management—mirroring the engineering perspective. Hence, it is further necessary to look at a community's risk for an event to happen, as well as the community's risk perception, i.e., how people respond to possible hazards (Ridzuan et al. 2018) as determinants of community resilience.

In fact, both risk management and risk perception have been developed as essential aspects of community resilience in theory as well as in the praxis. As demonstrated by Ridzuan et al. (2018), community risk perception is determined by community resilience elements such as community experience, community exposure, community reaction, community attitude, and community knowledge. Hence, communities that are, for example, informed and thus more aware of the risk are inherently more resilient. An analysis of community risk perception, just as much of actual risk, thus seems critical when looking to invite community action. A study or self-evaluation could thereby be applied. Withanachchi et al. (2018), for example, looked at the determinants of farmers' perceptions of water quality and risk and applied a mixed method approach including a household survey.

However, to be adequately resilient, the community most importantly must have both the resources available and the ability to apply or reorganize them in such a way to ensure essential functionality during and/or after the disturbance. Communities with a highly robust pool of resources and a high degree of adaptive capacity will be the most resilient. In contrast, when communities possess low levels of resources and low degrees of adaptive capacity, they will be less resilient. However, if a community is either high in resources or high in level of adaptive capacity, they can be relatively resilient if they take these assets into account in their planning. In that way, two communities could have an equal amount of resilience but a different mix of resources and adaptive capacity (Longstaff et al. 2010).

Besides risk perception, resource endowment, and adaptive capacity, other determinants of community resilience have been identified. Bec et al. (2016), for example, analyzed the key typologies of socio-ecological and community resilience that are evident within the literature. Besides structural stability, they include collaboration, leadership, and the ability to identify opportunities, or a positive outlook.

Which aspects are most relevant to each community, however, varies. As noted above, within any community, resilience is not evenly distributed since risk perception, access to resources, and adaptive capacity of individuals and groups are not either. Resilience depends critically on the socially differentiated capacities of different groups and individuals, including income, gender, or age. These capacities contribute to different vulnerabilities of social groups in cities to climate hazards, through features such as housing, location, and access to services or social networks (Moser and Satterthwaite 2010; Pelling 2003; Satterthwaite et al. 2009).

Some argue that the less of a role policy makers play, the more chance communities get to become resilient. This has been called the Darwinist approach as it leaves communities to fend for themselves. Others argue (e.g., Swanstrom 2008; Shaw 2012) that a diminished role for the state, despite indeed promoting rapid innovation, doesn't lead to increased resilience, but instead to disorganization and stress. Local authorities are needed for support and resources (Shaw and Maythorne 2012). However, equally it cannot be assumed that government is the primary guarantor of resilience, although it can be an important facilitator. In other words, community resilience cannot be directly controlled or engineered like a machine. But it can be fostered on multiple scales, realizing that different sectors and communities need different resilience-building actions (Swanstrom 2008).

1.1.5 Climate Adaptation

One of the major threats to SESs is climate change. Among other things, the second volume of the National Climate Assessment assigned some very specific costs to the economy of the USA if nothing is done to abate climate change. This includes US$118 billion from sea level rise; US$32 billion in infrastructure damage by the end of the century; and US$141 billion from heat-related deaths (NCA2018.Globalchange.gov 2019). Hence, climate change adaptation, or the ability to deal with climate-related disturbances, such as extreme weather events like hurricanes or heat waves, has become an important public policy domain.

It is now clear that the Earth's climate is changing and that, due to the inertia of the global climate system, it will not be possible to avoid all impacts even with the most drastic of GHG emissions reductions (Fünfgeld and McEvoy 2012). In fact, the 2018 report of the Intergovernmental Panel on Climate Change (IPCC), the United Nation's body for assessing the science related to climate change, predicts that the planet will reach the crucial threshold of 1.5 degrees Celsius (or 2.7 degrees Fahrenheit) as early as 2030, causing impacts such as more extreme weather, heat, and rising sea levels across the globe (IPCC.ch 2018).

Worldwide, 90% of large cities are located near coasts. In the USA, about 40% of all Americans live in coastal cities (Coast.noaa.gov 2019). Thus, billions of people and billions of dollars of built capital are directly at risk from sea level rise (SLR) and other climate-related risks. Cities can't afford not to invest in climate adaptation (The Nature Conservancy 2018). Climate adaptation measures are designed to protect existing assets, people, and places from the impact of climate variability and climate change.

However, when looking at current adaptation measures, it is evident that the engineering view on resilience (which assumes a return to an equilibrium or a steady-state after disturbance) prevails (Fünfgeld and McEvoy 2012), for example, as part of responding to global SLR. Here, building or augmenting physical infrastructure, such as sea walls and flood levees or beach nourishment, are common adaptation strategies. This corresponds with the idea of adaptation as an endpoint (Fünfgeld and McEvoy 2012), where communities or places become "more adapted" to climate change.

Overall, much adaptation action is framed as risk management. Fünfgeld and McEvoy (2012) interpret this as a pragmatic way of dealing with the uncertainty associated with the impact of climate change on complex SESs. As actions are mainly concerned with conserving the status quo and adopting a managerial, command-and-control understanding of systems, they are challenging to unite with the evolutionary, ever-changing nature of complex SESs. Moreover, some of the greatest stresses from climate change are likely to be indirect, incremental, or both (Tyler and Moench 2012). So, while these may be missed in vulnerability assessments and emergency planning, they can be managed through an evolutionary approach to resilience building.

As discussed above, however the standard approach in urban planning to increase resilience is through adjusting policies, practices, and plans in order to avoid negative impacts of climate change and to preserve the "status quo". This is known as adaptive management, or climate adaptation. In essence, this approach relies on prediction as the basis to identifying avenues for prevention (Tyler and Moench 2012). This approach comes with different challenges. First of all, predictions of anything, including future climate conditions, are uncertain as we can see in the abundance of different climate models and resulting prediction of consequences. Not the least as they are driven by competing agendas and underpinned by different assumptions in the first place. And even if planners can agree on future predictions to rely on, it may be difficult using them to identify likely impacts on climate change in their system (Milly et al. 2008). These "predict and prevent" approaches have also been criticized for their limited ability to deal with surprise (Wardekker et al. 2010), and their neglect of indirect effects, systemic weaknesses, or institutional constraints (e.g., Ericksen et al. 2009). Further, the adaptation planning approach tends to underemphasize the role of learning and governance as essential elements of ongoing adaptive management (Armitage et al. 2007). And finally, there is a lack of continuity in adaptive management since as different people are elected into political office, assumptions and priorities change, and so do actions on adaptation.

It is also important to note that climate change augments existing inequalities, rendering those most marginalized at greater peril to the health consequences of a changing climate (Watts et al. 2017). In fact, the first key message from the Lancet's Countdown on Climate Change and Health report emphasizes the disproportionate impact climate change has on the world's most marginalized people and the consequential impacts this has on these populations if social and environment justice concerns are not addressed. It states: "By undermining the social and environmental determinants that underpin good health, climate change exacerbates social, economic, and demographic inequalities, with the impacts eventually felt by all populations" (Watts et al. 2017).

Those who are at greatest risk to the effects of climate change are those who are most marginalized based on socially and environmentally mediated factors, such as socioeconomic status, culture, gender, race, employment, and education. Marginalized groups who tend to be the most affected by the mental and physical health implications of climate change are: Indigenous peoples, children, seniors, women, people with low-socioeconomic status, outdoor laborers, racialized people, immigrants, and people with preexisting health conditions (Watts et al. 2017). Consequently, any work on climate adaptation just like any work on resilience as argued earlier needs to consider power and politics (Swanstrom 2008) and be just and inclusive (C40.org). In practice, this can include, for example, stakeholder consultation in vulnerable communities before programs are designed. And during implementation, efforts should be made that programs have the capacity to meet individuals' where they are (in terms of knowledge and skills, for example) resources are equally accessible to all parts of the community.

1.2 Food Security and Climate Change

To conceptualize nutrition resilience, it seems useful to first look at food security, since nutrition is an often-overlooked part of it. A widely accepted definition of food security was developed at the World Food Summit in 1996. It states that:

> Food security exists when all people, at all times, have physical and economic access to sufficient, safe and nutritious food that meets their dietary needs and food preferences for an active and healthy life.

This definition reinforces the multidimensional nature of food security as described by the Food and Agricultural Organization of the United Nations (FAO)'s food security policy guideline (FAO.org 2006) and includes (based on Environmental Planning & Climate Protection Department 2014):

Food availability: The availability of sufficient quantities of food of appropriate quality, supplied through domestic production or imports (including food aid).
Food accessibility: Access by individuals to adequate resources (entitlements) for acquiring appropriate foods for a nutritious diet. This includes the affordability of food.
Utilization: Utilization of food through an adequate diet, together with clean water and health care to reach a state of nutritional well-being.
Stability: Access to adequate food at all times. Resilience to sudden shocks (e.g., economic crisis or severe weather), chronic stresses (e.g., heat, poverty rates), or cyclical events (e.g., drought) impact on food security stability.

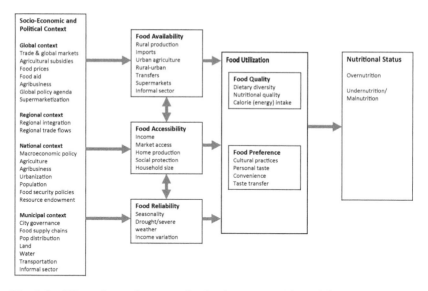

Fig. 1.1 Dimensions of community food security (Adapted from Frayne et al. 2010)

Figure 1.1 illustrates the dimensions of urban food security, which can be transferred to communities.

Looking at availability and accessibility, around 827 million people worldwide lacked access to adequate foods in 2017 (FAO.org 2018), and about one-third of the global population suffer from deficiencies in essential vitamins and minerals (WHO 2019). In 2016, around 108 million people worldwide were severely food insecure, a dramatic increase compared with 80 million in 2015 (UN Org 2017).

In the USA, about 48 million people (or 15% of the population) live in food insecure households. Almost 24 million people, or 8.4% of Americans are estimated to live in so-called food deserts (USDA 2010). Food deserts are defined as areas that lack access to affordable fruits, vegetables, whole grains, low-fat milk, and other foods that make up a full and healthy diet (Cdc.gov 2017). Most commonly they refer to urban areas with a large part of the population living more than one mile from a supermarket or large grocery store (in rural areas, the distance is more than 10 miles). Hence, the existence of food deserts is driven by whether there's a store, and whether community members possess the means to travel to it. Food access challenges affect primarily low-income neighborhoods. More than half of people living in food deserts are low income. And as of 2009, about 2.3 million Americans did not own a car and lived over one mile from a supermarket (Tulane University 2019).

A central challenge for nutritional resilience and component of food accessibility is food affordability. Several factors contribute to this challenge that fresh fruits and vegetables are generally more expensive than processed food. Firstly, farming broad acre crops like corn, soybean, or grains is inherently less expensive than growing fruits and vegetables. It is estimated that vegetables cost at least 10 times what grains cost to grow (*The Washington Post* 2017). On top of this inherent cost difference, crops that are more often used to produce unhealthy processed food (e.g., corn for syrups, soybean that acts as a filler in many processed foods) are subsidized by the US government, while fresh produce is not. This contributes to a higher cost of fruits and vegetables.

Additionally, arguments that people need to buy more expensive organic fruits and vegetables to eat healthily confuses consumers. This notion, which is supported by some influencers in the health and wellness space, is misguided as conventional produce that uses agrochemicals and sometimes relies on genetically modified seeds are heavily observed and regulated by the US Food and Drug Administration (FDA).

Aggarwal found that lower-income groups that prioritize nutrition during shopping are able to achieve diets comparable in quality to those of higher-income groups. According to Aggarwal et al. (2017), it is still possible to have a healthy and balanced diet even with low budgets. On the other hand, limited access to affordable, healthful foods is one factor that may make it harder for some Americans to eat a healthy diet. As illustrated in Fig. 1.1, food quality is part of the fourth dimension of food security: utilization. According to a National Center for Biotechnology Information article, studies show that people who live in neighborhoods where availability of healthy food is lowest are 55% less likely to enjoy good-quality diet compared to those who have better access to healthful food (Tulane University 2019).

The importance of being able to make the right choice when it comes to food becomes evident when looking at statistics. The fact that millions of people worldwide over-consume foods and beverages that are high in nutrients of concern, such as sugar and fat, has become an equally serious concern as eradicating hunger. Most of the world's population today live in countries where overweight and obesity kill more people than underweight (WHO.int 2018a). According to the WHO, in 2016 more than 1.9 billion adults worldwide were overweight. Of these, over 650 million were obese. In 2016, 41 million children under the age of 5 and over 340 million children and adolescents aged 5–19 were overweight or obese (WHO.int 2018a). In the USA, more than one-third of adults (39.8%, or over 93.9 million people) and 18.5% (or 13.7 million) of children and adolescents are obese. Estimates show that by 2030, almost half of Americans will be obese (USDA NIFA 2019).

Obesity is one of the leading causes of life-threatening diseases including cardiovascular diseases (mainly heart disease and stroke), musculoskeletal disorders and some cancers (including endometrial, breast, ovarian, prostate, liver, gallbladder, kidney, and colon) (WHO.int 2018a). Being overweight and obese, as well as these related non-communicable diseases, are largely preventable through adopting a health-promoting lifestyle that includes healthy nutrition. For example, it was found that people who live in neighborhoods where availability of healthy food is greater have a 45% reduced incidence of diabetes over a five-year period. The economic environment also affects health. Families that move to non-poor neighborhoods show a significantly reduced body mass index (BMI) (Tulane University 2019). Supportive environments and communities are fundamental in shaping people's choices, by making the choice of healthier foods and regular physical activity the easiest choice (the choice that is the most accessible, available, and affordable), and therefore preventing people from becoming overweight and obese (WHO.int 2018a).

Along with the impacts on societies' health and well-being, poor nutrition ultimately affects their economies. Worldwide, health costs both for the public and private sectors have been soaring, increasing at a faster pace than the economy of most countries (WHO.int 2018b). In the USA, obesity has resulted in costs of over US$147 billion for a single year (Finkelstein et al. 2009).

In addition to the costs, nutrition is linked to the long-term physical, emotional, societal, and economic well-being of those it affects and their communities. Physical as it can lead to a variety of lifestyle-related diseases. Emotional, since obesity and poor nutrition overall has been linked to a reduced mood and other mental health issues (Harvard Health Publishing 2015). Societal, since health challenges and early death disrupt families and communities. Poor nutrition also has been linked to developmental deficits in children (Korenman et al. 1995). And since these may prevent young adults to fulfill their potential and take advantage of opportunities, poor nutrition ultimately has an economic impact on communities.

Finally, when looking at urban food security, not only must one look at availability, accessibility, and affordability, but food preference needs to be considered. The healthy, nutritious food offered needs to be culturally relevant to the communities it's offered to, or it is not going to be consumed. In today's world where the social fabric of most cities includes immigrant cultures, cultural preferences and tastes must be accounted for. As such it becomes part of the utilization of food.

Of the dimensions of food security discussed above, climate change primarily threatens stability, as well as availability and accessibility. Climate change is predicted to cause a 3–4 degrees Celsius temperature rise and an increase in severe weather events worldwide (IPCC.ch 2018). It is expected to negatively affect existing levels of urban food security, and these are likely to fall disproportionately on the poor (Ziervogel and Frayne 2011). The focus on climate change impacts on food security has been primarily on food availability and production (Ziervogel and Ericksen 2010). It is however important to consider the effects of climate change on food accessibility, utilization, and stability (Table 1.1).

In the aftermath of a natural disaster, e.g., triggered by a warming climate, households that are already food insecure face additional challenges, while other households may become food insecure due to disaster-related expenses and hardships, such as loss of income or property damage (Zeuli and Nijhuis 2017).

1.3 Food Systems: The Vehicle for Community Nutrition Resilience

When looking at nutrition resilience, it is necessary to look at the resilience of the food system acting as the vehicle to supply sustainably produced and nutritious food to communities. As a sub-system of SESs, they constitute one of the core or critical systems of any city (Tyler and Moench 2012).

Food systems are described as dynamic interactions between and within the bio-geophysical and human environments that lead to the production, processing, preparation, and consumption of food (Ziervogel and Ericksen 2010). The dimensions and complexities

Table 1.1 Potential impacts of climate change on food systems and food security

Food system impacts	Food security impacts
Climate change scenario: increase in temperature	
Food production • Shift in agro-ecological zones • Change in crops grown per area • Decrease in yield due to heat stress • Increased weed pressure • Increased disease pressure • Heat stress impact on animal productivity • Reduction in fish number (coastal) *Food processing* • Increased need for cooling of perishable products • Change in postharvest losses *Food distribution* • Shorter shelf life of perishables • Improved refrigeration needed *Food consumption* • Food perishes quicker, requires more preservation or refrigeration	*Food availability* • Overall decrease in food supply • Shorter shelf life for perishable products reduces availability *Food accessibility* • Reduced availability leads to increase in food prices which would make food less affordable, particularly for urban populations *Food utilization* • Need to eat food sooner with shorter shelf life • Might require more fluid intake • Change in food types consumed *Food stability* • Reduction in stability of food supply due to decreased availability • Potential greater seasonal variation in supply • Potential greater seasonal variation in supply
Climate change scenario: increase in severe weather events, e.g., storms and floods	
Food production • Change in growing conditions (damaged crops, lower yields; soil erosion) • Impact on livestock health *Food processing* • Damaged storage facilities and processing plants *Food distribution* • Damage to transport network *Food consumption* • Food basket composition changed • Increased water-related health risks and cleanliness of food	*Food availability* • Decrease in food availability • Increased need for food aid • Increase in food imports *Food accessibility* • Increase in food prices make food less affordable • Food supply chains can be affected, resulting in allocation problems *Food utilization* • Food safety problems due to spoilage or emergency rations being used • Fresh, nutritious produce not available • Preferred foods not available *Food stability* • Overall decrease in food stability

Adapted from Ziervogel and Frayne (2011), Ziervogel and Ericksen (2010), and FAO (2008)

involved in food security and food systems illustrated in Fig. 1.1 indicate that food security on a local level cannot be divorced from global, regional, national, and municipal context.

According to Tyler and Moench (2012), the resilience of a food system is high when it is robust yet flexible and it can be accessed by high-capacity agents and where that access is enabled by supported institutions. There are different frameworks of food system resilience as well as different institutions advocating for the inclusion of food system resilience into resilience planning. Examples of these are discussed below.

The Milan Urban Food Policy Pact (Milanurbanfoodpolicypact.org 2017) is the most prominent global initiative that encourages city leaders to consider food systems in resilience planning, although it does so through a sustainability framework. Leaders from more than 180 cities worldwide, including nine US cities, have signed the Pact, pledging to work across government departments and food industry sectors to build resilient and sustainable food systems. The Food and Agriculture Organization (FAO) of the United Nations helped develop the framework of the Pact and works to support compliance, enable exchange of information and best practices between cities, and promote expansion of the program to more cities. While the Pact serves as a guiding force to help city leaders think about urban food systems, its focus on sustainability means that it does not fully consider the multidimensionality of resilience (Zeuli and Nijhuis 2017).

C40 is a network of more than 90 cities worldwide committed to addressing climate change. Its Food Systems Network (C40.org 2017), in partnership with EAT, a nonprofit founded by the Stordalen Foundation, Stockholm Resilience Centre, and the Wellcome Trust includes 26 cities that support efforts to reduce carbon emissions and increase resilience throughout their food systems. Its focus areas are food procurement, production, distribution, and waste however with a focus on mitigation.

The International Institute for Sustainable Development (IISD) is an independent think tank which works with governments, civil society, communities, and businesses to help them manage climate- and conflict-related risks, thus increasing resilience. They focus on participatory risk management and the identification and mitigation of conflict due to environmental change (IISD.org 2018).

The ICIC has developed an urban food system resilience framework that is oriented around a specific disaster that can affect a city and surfaces food system vulnerabilities to different types of natural and economic disasters (Zeuli and Nijhuis 2017). Although it doesn't address chronic stresses, it does make an analysis of food access at the neighborhood (=community) level possible. Thereby, the framework also surfaces specific areas (and populations) within the city that would be disproportionately impacted by food system disruptions.

As illustrated in Fig. 1.2, a food system comprises four main components involved with moving and transforming food from farm to table: food production, food processing, food distribution, and food access. Food production refers to all activities associated with growing crops and raising livestock. Food processing covers all aspects of cleaning, packaging, and processing at manufacturing facilities. Food distribution concerns the transport of food products from processing facilities to points of food access (e.g., grocery stores, restaurants, schools, and food banks). Food access is the level at which consumers can consume the food (Zeuli and Nijhuis 2017). The resilience of the system can be determined by looking at vulnerabilities at each different stage.

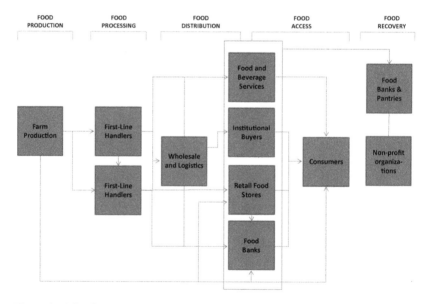

Fig. 1.2 The food system (Adapted from Zeuli and Nijhuis 2017)

Agriculture is one of the greatest contributors to climate change, and at the same time—and unlike other sectors that are large contributors such as transportation—it is the most vulnerable to changes in climate. According to the international research institute CGIAR, food systems contribute 19–29% of global anthropogenic GHG emissions. With the inclusion of food distribution and land use, its impact rises to 30%. In addition, worldwide more than 70% of deforestation in tropical and subtropical countries is driven by agriculture, exacerbating the issue of a warming atmosphere as carbon sinks disappear. Moreover, agriculture accounts for 50% of the land use and 80% of freshwater use in the USA (C40.org 2016).

Concurrently, climate change is impacting both crop production and causing rapid food and cereal price increases in many regions. At current rates of climate change, global food production is projected to decline by 2% every decade until 2050, just as the world's population is expected to reach 9.7 billion people. Hence, there may be scarcity of fruits, vegetables, and red meat. By the 2030s/2040s, between 40 and 80% of arable land used to grow staple crops like maize, millet, and sorghum could be lost due to the effects of higher temperatures, drought, and aridity (C40.org 2016).

Additionally, increased levels of carbon dioxide (CO_2) in the atmosphere are already decreasing the nutritional quality of crops—lowering their concentrations of vital nutrients like zinc and iron. The CO_2 levels in the second half of this century would likely reduce the levels of zinc, iron, and protein in wheat, rice, peas, and soybeans even more (Zhu et al. 2018).

Vulnerabilities in **food processing** are typically not a large concern for local food availability. This is because the majority of food is not processed and packaged locally (Zeuli and Nijhuis 2017). Exceptions are highly perishable goods, like milk, that need to be transported from farm to consumer relatively quickly. That's why typically cities have milk processing plants locally. Furthermore, in many cities, local food production (urban farming) and local food processing are expanding, partly to mitigate risks associated with importing food grown in areas susceptible to climate change, e.g., drought in California.

Vulnerabilities in **food distribution** exist due to the complex systems with different routes of getting food from farm to consumers. The vulnerabilities are mostly centered around food retail supply chains as this is where communities get their food. Additionally, food retailers can have three different types of suppliers, which also contribute to distribution vulnerability. These supplier types are: a primary warehouse supplier (wholesaler or distribution center); secondary suppliers that source products the primary supplier doesn't carry (specialty products) and; direct-to-store delivery for limited portfolio of products such as prepared frozen food, milk, and some fresh produce. Most supermarkets usually own their primary supplier, while smaller stores rely on independently owned warehouse suppliers (Zeuli and Nijhuis 2017).

Overall, food distribution, a very consolidated and competitive industry, has gone to great lengths to minimize its own vulnerabilities. Warehouse suppliers will do everything to recover from natural disasters as quickly as possible and deliver supplies to avoid losing customers. Many large food retailers even require their warehouse suppliers to have business continuity plans in place to secure their purchasing contracts. Food distribution networks are generally fragmented with different suppliers in different locations, creating some resilience in the system. National suppliers are also well prepared to handle disruptions due to their multiple locations in different states and resources to invest in structural improvements to withstand disasters (Zeuli and Nijhuis 2017). Therefore, the primary vulnerabilities of food distribution are the location of warehouses and suppliers and the reliance on primary transportation routes (roads, bridges, tunnels), since nearly all food is distributed by truck (Zeuli and Nijhuis 2017).

In regard to **food access, particularly retail** vulnerabilities, there are three factors that determine food system resilience: the number of retail stores per capita; the mix of supermarkets, grocery stores, and convenience stores; and the location of food retail stores in "at risk" areas (Zeuli and Nijhuis 2017). Supermarkets are defined as grocery stores with US$2 million or more in annual revenue. A grocery store is a retail store that sells a variety of food products, including some perishable items and general merchandise. And a convenience store is defined as a small, easy-access food store with a limited assortment per the definition used by the Food Marketing Institute (FMI.org 2019).

In addition, an examination of the food system resilience should consider food consumption at institutions (e.g., schools, hospitals, prisons, caterers, restaurants). Most institutions are supplied by national food service providers and are less vulnerable to local disasters (Zeuli and Nijhuis 2017). However, when schools are shut after a disaster, this exacerbates the pressure on families to feed kids at home which may be extremely challenging for low-income families.

Food banks are the backbone of urban food safety nets (Zeuli and Nijhuis 2017). Their role is to distribute donated and purchased groceries directly to food insecure families. However, it has been found that they are limited in their capacity to improve overall food security outcomes due to the limited provision of nutrient-dense foods in insufficient amounts, especially from vegetables and fruits. Food banks have the potential to improve food security outcomes when operational resources are adequate, provisions of perishable food groups are available, and client needs are identified and addressed (Bazerghi et al. 2016).

Across the food system, another challenge that affects resilience is its efficiency. Currently, about one-third of all food is wasted. In developing countries, 40% of losses occur at post-harvest and processing levels while in industrialized countries such as the USA more than 40% of losses happen at retail and consumer levels (FAO.org 2019a). Fruits and vegetables thereby have the highest wastage rates of any food. This is a strong signal that the food system is not efficient, particularly at access level with regard to healthy food choices.

Food recovery, and the avoidance of food waste, has been named one of the top three opportunities to reduce GHG emissions globally (Drawdown 2019). Worldwide the energy used for the production, harvesting, transporting, and packaging of wasted food generates more than 3.3 billion metric tons of carbon dioxide. In addition, as food decomposes, it creates methane gas, which is at least 28 times more potent than carbon dioxide (United States Environmental Protection Agency 2019). According to the UN Food and Agriculture Organization (FAO.org), 8% of all global GHG come from food waste. If food waste were a country, it would be the world's third largest emitter of GHG, behind the USA and China (FAO.org 2019b). According to the FAO, the land devoted to producing wasted food is roughly 5.4 million square miles, which would make it the second largest country in the world behind Russia. Moreover, recovered food can alleviate hunger and even nutritional challenges of underprivileged communities, if distributed in-time.

Overall, it appears that food systems have been largely overlooked in cities' resilience planning efforts in the USA. Most cities expect to provide residents with food for a relatively short period of time—a few weeks at most—during the immediate aftermath of a natural disaster. But as Hurricane Katrina in New Orleans or Hurricane Irma in Miami demonstrated, food system interruptions may last months or years. In addition, as previously discussed, resilience within a system, such as a city, is oftentimes unequally distributed; hence, long-lasting disruptions can create significant food access issues. This is especially the case for populations that are already food insecure (Zeuli and Nijhuis 2017).

Zeuli and Nijhuis (2017) offer some ideas on why food systems have not yet been in focus for the resilience planning in many cities. City leaders may assume that resilience planning for the city's infrastructure in general is sufficient, since food distribution and retail depend on transportation and utilities. However, these are not the only parts that matter. In addition, city leaders may assume that since food systems are predominantly composed of private-sector businesses, food businesses have sufficient resources and motive to rapidly return to normal operations. However, this may not always be true. Smaller grocery stores and corner stores are less resilient for a variety of reasons. For example, as they're not part of a national chain, they may face longer periods of closure after a disturbance because they have fewer resources and are less likely to have sufficient insurance or business continuity plans. Some businesses simply don't have the resources to reopen while they are waiting for their insurance payment to come through. Also, the application process for public disaster recovery funds often requires a lot of time, and the distribution of funds is often delayed and inefficient. Furthermore, there exist no incentives to invest in making space more disaster-resistant, as it's rented. In addition, there are uncertainties with regard to demand particularly shortly after a disaster, due to the financial hardship of communities, and since telecommunications networks might not be working so that benefits from the Federal Supplemental Nutrition Assistance Program (SNAP) can't be processed. Lastly, food safety regulations may prohibit stores from opening before they have been inspected, and that waiting period without revenues may put business owners out of business.

True resilience thereby would include resilience against evolutionary changes such as an aging population and immigration as well as against disruptions such as economic or climatic shocks (Zeuli and Nijhuis 2017). Cities that focus on developing resilient food systems will ensure that (1) there are no systemic challenges affecting the steps of the food system, and (2) that food supplies return to pre-disaster levels as quickly and as equitably as possible so that all residents have adequate access to food in their neighborhoods (Zeuli and Nijhuis 2017).

1.4 Community Nutrition Resilience: Moving from "Survive" to "Thrive"

For the reasons discussed above, the term community food security hence insufficiently describes the overarching goal. Focus should be on building and maintaining resilience as in the ability of the community to maintain a certain level of food security, being able to bounce back from shocks and stresses and evolve and progress over time. Furthermore, in developed countries such as the USA, the primary goal of food system resilience should not merely be food security, as in the existence of access, availability, and affordability, but it should be on quality, i.e., the nutritional dimensions of food security. Cultural relevance thereby plays a role in ensuring uptake of healthy, nutritious food as discussed.

It should be the ultimate goals of any community-related planning and programming to make sure that communities have access to, and can afford, nutritious food. This will ultimately provide them with the opportunity to recoup the physical, mental, and emotional benefits of being nourished properly, longevity, and optimal health, and hence not only survive but thrive.

Considering an apparent lack of a definition of community nutrition resilience in literature, this book offers the following:

> Community nutrition resilience is the ability of a community (such as a neighborhood or otherwise locally confined SES) to deal with sudden shocks or adapt to slowly progressing changes that affect its nutritional status. It ensures that there is availability of, access to, and affordability of nutritious food that provides community members with the optimal basis for physical, mental, emotional and social development.

The following chapter will look at the chronic stresses and acute shocks threatening community nutrition resilience in Greater Miami.

References

100ResilientCities.org. (2018). *Frequently asked questions (FAQ) about 100 resilient cities* [online]. Available at: http://www.100resilientcities.org/100RC-FAQ/#/-_/. Accessed 29 January 2018.

Adger, W. N. (2003). Building resilience to promote sustainability. *IHDP Update* (pp. 1–3).

Aggarwal, A., Rehm, C. D., Monsivais, P., & Drewnowski, A. (2017). *Importance of taste, nutrition, cost and convenience in relation to diet quality: Evidence of nutrition resilience among US adults using National Health and Nutrition Examination Survey (NHANES) 2007–2010*. Prev. Med. Authormanuscript; available in PMC 2017 September 01. 90: 184–192. https://doi.org/10.1016/j.ypmed.2016.06.030.

Armitage, D., Berkes, F., & Doubleday, N. (2007). Introduction: Moving beyond co-management. In D. Armitage, F. Berkes, & N. Doubleday (Eds.), *Adaptive co-management: Collaboration, learning and multi-level governance* (pp. 1–15). Vancouver and Toronto: UBC Press.

Bazerghi, C., McKay, F. H., & Dunn, M. (2016). The role of food banks in addressing food insecurity: A systematic review. *Journal of Community Health, 41*(4), 732–740. https://doi.org/10.1007/s10900-015-0147-5.

Bec, A., McLennan, C. L., & Moyle, B. D. (2016). Community resilience to long-term tourism decline and rejuvenation: A literature review and conceptual model. *Current Issues in Tourism, 19*(5), 431–457.

Berkes, F., & Folke, C. (1998). *Linking social and ecological systems: Management practices and social mechanisms for building resilience*. Cambridge: Cambridge University Press.

Brundtland, G. (1987). *Report of the world commission on environment and development: Our common future* [online]. Available at: http://www.un-documents.net/our-common-future.pdf. Accessed 6 September 2018.

C40.org. (2016). *C40's executive director Mark Watts: Mayors are changing the way we think about food* [online]. Available at: https://www.c40.org/blog_posts/c40-s-executive-director-mark-watts-mayors-are-changing-the-way-we-think-about-food. Accessed 20 September 2018.

C40.org. (2017). *Urban planning and development initiative: Food systems* [online]. Available at: http://www.c40.org/networks/food_systems. Accessed 22 January 2018.

Carpenter, S. R., Westley, F., & Turner, G. (2005). Surrogates for resilience of social-ecological systems. *Ecosystems, 8*(8), 941–944. https://doi.org/10.1007/s10021-005-0170-y.

Cdc.gov. (2017). *Food desert* [online]. Available at: https://www.cdc.gov/healthcommunication/toolstemplates/entertainmented/tips/FoodDesert.html. Accessed 16 December 2018.

Coast.noaa.gov. (2019). *Fast facts: Economics and demographics* [online]. Available at: https://coast.noaa.gov/states/fast-facts/economics-and-demographics.html. Accessed 1 March 2019.

Davoudi, S. (2012). Resilience: A bridging concept or a dead end? In S. Davoudi & L. Porter (Eds.), Applying the resilience perspective to planning: Critical thoughts from theory and practice. *Planning Theory & Practice, 13*(2), 299-307. https://doi.org/10.1080/14649357.2012.677124.

Drawdown. (2019). *Food: Reduced food waste* [online]. Available at: https://www.drawdown.org/solutions/food/reduced-food-waste. Accessed 12 February 2019.

Environmental Planning & Climate Protection Department. (2014). *Durban climate change strategy: Food security theme report: Draft for public comment* [online]. Available at: http://www.durban.gov.za/City_Services/energy-office/Documents/DCCS%20Food%20Security%20Theme%20Report.pdf. Accessed 15 January 2019.

Ericksen, P. J., Ingram, J. S. I., & Liverman, D. M. (2009). Food security and global environmental change: Emerging challenges. *Environmental Science & Policy, 12*, 373–377. https://doi.org/10.1016/j.envsci.2009.04.007.

FAO. (2008). *Climate change adaptation and mitigation in the food and agriculture sector.* Technical Background Document from the Expert Consultation held on 5 to 7 March 2008, FAO, Rome.

FAO.org. (2006). *Food security: Policy brief* [online]. Available at: http://www.fao.org/forestry/13128-0e6f36f27e0091055bec28ebe830f46b3.pdf. Accessed 22 January 2018.

FAO.org. (2018). *The state of food security and nutrition in the world* [online]. Available at: http://www.fao.org/state-of-food-security-nutrition/en/. Accessed 17 December 2018.

FAO.org. (2019a). *Key facts on food loss and waste you should know!* [online]. Available at: http://www.fao.org/save-food/resources/keyfindings/en/. Accessed 2 February 2019.

FAO.org. (2019b). *Food wastage footprint & climate change* [online]. Available at: http://www.fao.org/3/a-bb144e.pdf. Accessed 1 March 2019.

Finkelstein, E. A., Trogdon, J. G., Cohen, J. W., & Dietz, W. (2009). Annual medical spending attributable to obesity: Payer-and service-specific estimates. *Health Affairs (Millwood), 28*(5), 822–831. https://doi.org/10.1377/hlthaff.28.5.w822.

FMI.org. (2019). *Food industry glossary* [online]. Available at: https://www.fmi.org/our-research/food-industry-glossary. Accessed 13 February 2019.

Folke, C. (2006). Resilience: The emergence of a perspective for social–ecological systems analyses. *Global Environmental Change, 16*(3), 253–267.

Folke, C., Carpenter, S., Walker, B., Scheffer, M., Chapin, T., & Rockstrom, J. (2010). Resilience thinking: Integrating resilience, adaptability and transformability. *Ecology and Society, 15*(4), 20–28. https://doi.org/10.5751/ES-03610-150420.

Frayne, B., Pendleton, W., Crush, J., Acquah, B., Battersby-Lennard, J., Bras, E., et al. (2010). *The state of urban food insecurity in Southern Africa.* Urban Food Security Series No. 2. Kingston and Cape Town: Queen's University and AFSUN.

Fünfgeld, H., & McEvoy, D. (2012). *Framing climate change adaptation in policy and practice* (Working Paper 1). Melbourne: Victorian Center for Climate

Change Adaptation Research. Available at: http://vcccar.org.au/files/vcccar/ Framing_project_workingpaper1_190411.pdf. Accessed 8 February 2012.

Harvard Health Publishing. (2015). *Nutritional psychiatry: Your brain on food* [online]. Available at: https://www.health.harvard.edu/blog/nutritional-psychiatry-your-brain-on-food-201511168626. Accessed 10 February 2019.

Hodgson, G. M. (2006). What are institutions? *Journal of Economic Issues, 40*(1), 1–25. https://doi.org/10.1177/0170840607067832.

Holling, C. S. (1973). Resilience and stability of ecological systems. *Annual Review of Ecology and Systematics, 4*, 1–23. https://doi.org/10.1146/annurev.es.04.110173.000245.

Holling, C. S. (1996). Engineering resilience versus ecological resilience. In P. Schulze (Ed.), *Engineering within ecological constraints*, p. 32. Washington, DC: The National Academies Press.

Holling, C. S., & Gunderson, L. H. (2002). Resilience and adaptive cycles. In L. H. Gunderson & C. S. Holling (Eds.), *Panarchy: Understanding transformations in human and natural systems* (pp. 25–62). Washington, DC: Island Press.

Huntjens, P., Lebel, L., Pahl-Wostl, C., Camkin, J., Schulze, R., & Kranz, N. (2012). Institutional design propositions for the governance of adaptation to climate change in the water sector. *Global Environmental Change, 22*, 67–81. https://doi.org/10.1016/j.gloenvcha.2011.09.015.

IISD.org. (2018). *Resilience* [online]. Available at: https://www.iisd.org/program/resilience. Accessed 13 February 2019.

IPCC.ch. (2018). *Summary for policymakers of IPCC special report on global warming of 1.5°C approved by governments* [online]. Available at: https://www.ipcc.ch/2018/10/08/summary-for-policymakers-of-ipcc-special-report-on-global-warming-of-1-5c-approved-by-governments/. Accessed 16 December 2018.

Kim, T. J., & von dem Knesebeck, O. (2018). Income and obesity: What is the direction of the relationship? A systematic review and meta-analysis. *BMJ Open, 8*(1), e019862. https://doi.org/10.1136/bmjopen-2017-019862.

Kinzig, A. P., Ryan, P., Etienne, M., Allison, H., Elmqvist, T., & Walker, B. H. (2006). Resilience and regime shifts: Assessing cascading effects. *Ecology and Society, 11*(1), 20.

Korenman, S., Miller, J. E., & Siaastad, J. E. (1995). Long-term poverty and child development in the United States: Results from the NLSY. *Children and Youth Services Review, 17*(1–2), 127–155. https://doi.org/10.1016/0190-7409(95)00006-x.

Little, R. G. (2002). Controlling cascading failure: Understanding the vulnerabilities of interconnected infrastructures. *Journal of Urban Technology, 9*(1), 109–123. https://doi.org/10.1080/106307302317379855.

Longstaff, P. H., Armstrong, N. J., Perrin, K., Parker, W. M., & Hidek, M. A. (2010). Building resilient communities: A preliminary framework for assessment. *Homeland Security Affairs, VI*(3), 1–23.

Milanurbanfoodpolicypact.org. (2017). *Milan urban food policy pact* [online]. Available at: http://www.milanurbanfoodpolicypact.org/. Accessed 2 February 2019.
Milly, P. C. D., Betancourt, J., Falkenmark, M., Hirsch, R. M., Kundzewicz, Z. W., Lettenmaier, D. P., et al. (2008). Stationarity is dead: Whither water management? *Science, 319*(5863), 573–574. https://doi.org/10.1126/science.1151915.
Moser, C., & Satterthwaite, D. (2010). Towards pro-poor adaptation to climate change in the urban centers of low- and middle-income countries. In R. Mearns & A. Norton (Eds.), *Social dimensions of climate change: Equity and vulnerability in a warming world* (pp. 231–258). Washington, DC: World Bank.
NCA2018.Globalchange.gov. (2019). *Fourth national climate assessment* [online]. Available at: https://nca2018.globalchange.gov. Accessed 1 February 2019.
Norris, F. H., Stevens, S. P., Pfefferbaum, B., Wyche, K. F., & Pfefferbaum, R. L. (2008). Community resilience as a metaphor, theory, set of capabilities, and strategy for disaster readiness. *American Journal of Community Psychology, 41*, 127–150.
Olsson, P., Gunderson, L. H., Carpenter, S., Ryan, P., Lebel, L., Folke, C., et al. (2006). Shooting the rapids: Navigating transitions to adaptive governance of social-ecological systems. *Ecology and Society, 11*(1), 18.
Pelling, M. (2003). *The vulnerability of cities: Natural disasters and social resilience*. London: Earthscan.
Porter, L., & Davoudi, S. (2012). The politics of resilience for planning: A cautionary note. In S. Davoudi, K. Shaw, L. J. Haider, A. E. Quinlan, A. E., G. D. Peterson, C. Wilkinson, H. Fünfgeld, D. McEvoy, & L. Porter (Eds.), Resilience: A bridging concept or a dead end? "Reframing" resilience: Challenges for planning theory and practice interacting traps: Resilience assessment of a pasture management system in Northern Afghanistan urban resilience: What does it mean in planning practice? Resilience as a useful concept for climate change adaptation? The politics of resilience for planning: A cautionary note. *Planning Theory & Practice, 13*(2), 324–328. https://doi.org/10.1080/14649357.2012.677124.
Quinlan, A. (2003). Resilience and adaptive capacity: Key components of a sustainable social-ecological system. *IHDP Update, 2*, 4–5.
Resalliance.org. (2018). *Resilience* [online]. Available at: https://www.resalliance.org/resilience. Accessed 16 December 2018.
Resilience Alliance. (2010). *Assessing resilience in social-ecological systems: A workbook for practitioners, version 2.0* [online]. Available at: http://www.resalliance.org/3871.php. Accessed 27 February 2018.
Ridzuan, A. A., Oktari, R. S., Zainol, N. A. M., Abdullah, H., Liaw, J. O. H., Mohaiyadin, N. M. H., et al. (2018). *Community resilience elements and community risk perception at Banda Aceh province, Aceh, Indonesia*. MATEC Web of Conferences (Vol. 229, p. 01005). EDP Sciences.

Satterthwaite, D., Dodman, D., & Bicknell, J. (2009). Conclusions: Local development and adaptation. In J. Bicknell, D. Dodman, & D. Satterthwaite (Eds.), *Adapting cities to climate change: Understanding and addressing the development challenges* (pp. 359–383). London: Earthscan.

Scheffer, M. (2009). *Critical transitions in nature and society*. Princeton, NJ: Princeton University Press.

Seville, E. (2009). Resilience: Great concept…but what does it mean for organisations? In Ministry of Civil Defence & Emergency Management (Ed.), *Community resilience: Research, planning and civil defence emergency management*. Wellington, NZ: Ministry of Civil Defence & Emergency Management.

Shaw, K. (2012). "Reframing" resilience: Challenges for planning theory and practice. In S. Davoudi & L. Porter (Eds.), Applying the resilience perspective to planning: Critical thoughts from theory and practice. *Planning Theory & Practice, 13*(2), 308–312. https://doi.org/10.1080/14649357.2012.677124.

Shaw, K., & Maythorne, L. (2012). Managing for local resilience: Towards a strategic approach. *Public Policy and Administration, 28*(1), 43–65.

Swanstrom, T. (2008). *Regional resilience: A critical examination of the ecological framework* (Working Paper, 2008[7]). Berkeley: University of California, Institute of Urban and Regional Development.

The Nature Conservancy. (2018). *Climate action from the ground up* [online]. Available at: https://global.nature.org/content/climate-action-from-the-ground-up?src3=e.gp.nat.Sept2018.National.readmore. Accessed 20 September 2018.

The Washington Post. (2017). *Junk food is cheap and healthful food is expensive, but don't blame the farm bill* [online]. Available at: https://www.washingtonpost.com/lifestyle/food/im-a-fan-of-michael-pollan-but-on-one-food-policy-argument-hes-wrong/2017/12/04/c71881ca-d6cd-11e7-b62d-d9345ced896d_story.html?utm_term=.467a9e106b47. Accessed 10 May 2019.

Tulane University. (2019). *Food deserts in America (infographic)* [online]. Available at: https://socialwork.tulane.edu/blog/food-deserts-in-america. Accessed 2 February 2019.

Tyler, S., & Moench, M. (2012). A framework for urban climate resilience. *Climate and Development, 4*(4), 311–326. https://doi.org/10.1080/17565529.2012.745389.

United States Environmental Protection Agency. (2019). *Sustainable management of food basics* [online]. Available at: https://www.epa.gov/sustainable-management-food/sustainable-management-food-basics. Accessed 2 February 2019.

UN Org. (2017). *UN, partners warn 108 million people face severe food insecurity worldwide* [online]. Available at: http://www.un.org/apps/news/story.asp?NewsID=56472#.WmZFu5M-cWo. Accessed 22 January 2018.

USDA. (2010). *Access to affordable, nutritious food is limited in "Food Deserts"* [online]. Available at: https://www.ers.usda.gov/amber-waves/2010/march/access-to-affordable-nutritious-food-is-limited-in-food-deserts/. Accessed 20 December 2018.

USDA NIFA. (2019). *Obesity* [online]. Available at: https://nifa.usda.gov/topic/obesity. Accessed 2 February 2019.

Vale, L. J., & Campanella, T. J. (2005). *The resilient city: How modern cities recover from disaster.* New York: Oxford University Press.

Walker, B., Holling, C. S., Carpenter, S., & Kinzig, A. (2004). Resilience, adaptability and transformability in social-ecological systems. *Ecology and Society, 9*(2), 5.

Wardekker, J. A., de Jong, A., Knopp, J. M., & van der Sluijs, J. P. (2010). Operationalizing a resilience approach to adapting a delta to uncertain climate changes. *Technological Forecasting and Social Change, 77,* 987–998. https://doi.org/10.1016/j.techfore.2009.11.005.

Watts, N., Amann, M., Ayeb-Karlsson, S., Belesova, K., Bouley, T., Boykoff, M., et al. (2017). The Lancet countdown on health and climate change: from 25 years of inaction to a global transformation for public health. *Lancet, 391*(10120), 581–630. https://doi.org/10.1016/S0140-6736(17)32464-9.

Whalen, R., & Zeuli, K. (2017). *Resilient cities require resilient food systems.* The Rockefeller Foundation [online]. Available at: https://www.rockefellerfoundation.org/blog/resilient-cities-require-resilient-food-systems/. Accessed 18 December 2017.

WHO.int. (2018a). *Obesity and overweight: Key facts* [online]. Available at: https://www.who.int/news-room/fact-sheets/detail/obesity-and-overweight. Accessed 16 December 2018.

WHO.int (2018b). *Public spending on health: A closer look at global trends* [online]. Available at: https://www.who.int/health_financing/documents/health-expenditure-report-2018/en/. Accessed 16 December 2018.

Wilkinson, C. (2012). Urban resilience: What does it mean in planning practice? In S. Davoudi, K. Shaw, L. J. Haider, A. E. Quinlan, G. D. Peterson, C. Wilkinson, H. Fünfgeld, D. McEvoy, L. Porter, & L. Porter (Eds.), Resilience: A bridging concept or a dead end? "Reframing" resilience: Challenges for planning theory and practice interacting traps: Resilience assessment of a pasture management system in Northern Afghanistan urban resilience: What does it mean in planning practice? Resilience as a useful concept for climate change adaptation? The politics of resilience for planning: A cautionary note. *Planning Theory & Practice, 13*(2), 324–328. https://doi.org/10.1080/14649357.2012.677124.

Wilkinson, C., Porter, L., & Colding, J. (2010). Metropolitan planning and resilience thinking: A practitioner's perspective. *Critical Planning, 17*(17), 25–44.

Withanachchi, S., Kunchulia, I., Ghambashidze, G., Al Sidawi, R., Urushadze, T., & Ploeger, A. (2018). Farmers' perception of water quality and risks in the Mashavera River Basin, Georgia: Analyzing the vulnerability of the social-ecological system through community perceptions. *Sustainability, 10*(9), 3062.

Zeuli, K., & Nijhuis, A. (2017). *The resilience of America's urban food systems: Evidence from five cities* [ebook]. Roxbury, MA: ICIC. Available at: http://

icic.org/wp-content/uploads/2017/01/Rockefeller_ResilientFoodSystems_FINAL_post.pdf?x96880. Accessed 18 December 2017.

Zhu, C., Kobayashi, K., Loladze, I., Zhu, J., Jiang, Q., Xu, X., et al. (2018). Carbon dioxide (CO_2) levels this century will alter the protein, micronutrients, and vitamin content of rice grains with potential health consequences for the poorest rice-dependent countries. *Science Advances*, 4(5), eaaq1012. https://doi.org/10.1126/sciadv.aaq1012.

Ziervogel, G., & Ericksen, P. J. (2010). *Adapting to climate change to sustain food security: Advanced review* [online]. Available at: wires.wiley.com/climatechange.

Ziervogel, G., & Frayne, B. (2011). *Climate change and food security in Southern African cities* (Urban Food Security Series No. 8). Kingston and Cape Town: Queen's University and AFSUN.

CHAPTER 2

Resilience Challenges to Community Nutrition Security in Greater Miami

Abstract This chapter examines the challenges affecting Greater Miami's environmental, social, and economic resilience. It looks at influencing factors and historical developments shaping the region, such as political environment, immigration, and events triggered by a changing climate. It then discusses the status quo of community nutrition resilience in Greater Miami based on selected food security and health parameters. It also examines in detail the main threats to community nutrition resilience in Greater Miami. These include a lack of focus, policy, and planning, as well as threats related to the steps of the food system and particularly the prevalent lack of health literacy. Finally, it introduces the subsequent chapters of the book.

Keywords Miami · Historical analysis · Climate change · Resilience challenges · Community nutrition resilience

The Miami Metropolitan area, also known as Greater Miami and as defined by the Office of Management and Budget (Whitehouse.gov 2018), consists of the three counties Miami-Dade (MDC), Broward, and Palm Beach and is the seventh largest metropolitan area in the USA. With 6.2 million people living in Greater Miami, there is a growing concern for livability and growth should they continue to see the same stresses and shocks they've seen in the past three decades.

© The Author(s) 2020
F. Alesso-Bendisch, *Community Nutrition Resilience in Greater Miami*, Palgrave Studies in Climate Resilient Societies,
https://doi.org/10.1007/978-3-030-27451-1_2

Rapid growth has transformed Florida's landscape from rural to urban in a few decades. By 1990, 85% of its people lived in urban areas. Swamps have become golf courses and shopping malls. Once-isolated beaches are now dominated by high-rise hotels and condominiums. Cities have arisen with both skyscrapers and slums, surrounded by burgeoning suburbs with their own schools, hospitals, and shopping centers.

A growing population, an aging infrastructure, decreasing housing quality and affordability are the biggest stressors to Miami's resilience. These are being exacerbated increasingly by incremental environmental hazards due to climate change, such as rising sea levels, as well as by shocks such as floods and hurricanes. All of these leave Greater Miami's communities increasingly vulnerable.

Greater Miami is vulnerable to hurricanes and tropical storms, particularly during the hurricane season from June 1 through November 30, and this vulnerability is expected to increase in future. The National Climate Assessment (NCA2018.Globalchange.gov 2019) predicts hotter days, more hurricanes, more outbreaks of insect-borne disease, greater SLR impacts, and more economic damage from climate change in the coming years. Increased development and population over the last 11 years has created uncertainties of evacuation, response, and long-term community recovery. During and shortly after storms, power availability, communication, and access to residents are key. But residents and governments lack funding for storm preparedness and absentee owners contribute to delayed recovery (Resilient305.org 2018).

Behind California, Florida, is the second most vulnerable state for climate change, dubbed "Ground Zero". Natural disasters and climate change deliver hard-hitting blows to the economy of Greater Miami in ways such as the inability to accommodate tourism and damage or devastation caused to the highly valued agricultural land. Additionally, those already living in poverty, struggle even more to access food, clean water, and experience further displacement from the population. Within the local communities, a lack of political will, adequate funding, partnerships, and collaborations has limited the resources necessary to deal with these challenges. At this point, with the impending threats of natural shocks to come, it is imperative that the region's leaders focus on taking actions to plan and prepare for Greater Miami's resilience.

Resilience in the urban context has been defined by the Rockefeller Foundation's 100RC, and adapted by Greater Miami, as the "capacity of individuals, communities, institutions, businesses, and systems within a city to survive, adapt, and grow no matter what kinds of chronic stresses and acute shocks they experience" (100ResilientCities.org 2018).

This definition, also adopted by this book, stresses two equally important foci on resilience: first, the challenge of maintaining resilience in the face of chronic stresses (e.g., urbanization, population aging, immigration) and secondly, resilience when faced with acute shocks such as severe weather events or sudden crises. It thereby advocates an evolutionary perspective on resilience, meaning that resilience is not conceived as the ability to return to normality, but rather as the ability to change, adapt, and transform in response to stresses and strains (Carpenter et al. 2005).

This chapter will look at the different challenges Greater Miami is facing that are threatening the resilience of its communities. It will focus on nutrition-related challenges, but also look at challenges that indirectly affect food availability, access, and stability. As discussed in Chapter 1, communities are resilient if they both have the resources available and the ability to apply them, i.e., adaptive capacity (Longstaff et al., 2010). An analysis of both of these parameters will be center to this discussion.

2.1 HISTORICAL DYNAMICS

Although a full and thorough resilience assessment cannot be conducted for Greater Miami within the scope of this book, a basic resilience analysis will be attempted by following the resilience assessment workbook method proposed by the Resilience Alliance. This suggests that practitioners should first look at *"how historical system dynamics have shaped the current system. Social-ecological systems are dynamic and…having a broad overview of system change through time can reveal system drivers, the effects of interventions, past disturbances and responses"* (Resilience Alliance 2007, p. 22).

2.1.1 Political Support

Possibly, the most influencing factor related to this discussion is the political support for resilience efforts in Greater Miami. This support can be provided on federal, state and local level, all of which are discussed below.

On the federal level, political support has been changing with the Administrations. In 2008 under President Obama, there were grants available to regions to create sustainable communities, encouraging them to look at climate adaptation. The Southeast Florida Regional Climate Change Compact, one of the main collaborative efforts to tackle climate resilience in Greater Miami, was partly funded through these grants (the Compact will be discussed in detail in Chapter 3 of this book). Superstorm Sandy in 2012, the deadliest and most destructive hurricane of the 2012 Atlantic hurricane season inflicted almost US$70 billion in damages nationwide, motivated Congress to allocate budget for recovery. President Obama was then able to fund committees that addressed resilience. At that time, the Rockefeller Foundation started defining resilience, which ultimately led to the creation of 100RC.

When looking at politics in the State of Florida, they have traditionally been conflicted between its liberal Southeastern region and its conservative northwestern region. Thus, Florida as a "swing state" could reasonably be won by either the Democratic or Republican presidential candidate. Opposing views in politics often revolve around budgeting and how money for budgets should be raised and hence the support for resilience efforts, including the response to climate change is no exception. On the contrary, particularly the latter has historically seen massive ups and downs, depending on the federal and state government at a time.

Already Gov. Crist, a Conservative Republican who governed between 2007 and 2011, focused on climate change in his tenure, which by some has been dubbed Florida's high point in focus on climate change. The American Recovery and Reinvestment Act of 2009, a stimulus package to combat the recession, provided funding to work on renewable energy, and the Florida Energy and Climate Commission were formed. This was one of the first tools to provide Greater Miami with resources to invest in climate infrastructure. Governor Christ also encouraged the launch of the Compact.

Until 2018, Florida was governed by Gov. Rick Scott, a traditional Republican, who has refused to publicly address climate change declaring that he is not a scientist. Reportedly, he has banned the phrase in his administration (a charge he denies) and backed up President Donald Trump's decision to withdraw from the Paris accord (Miami Herald 2018d). During Scott's administration, support for sustainability and resilience work from state level receded and several initiatives introduced by Gov. Christ were terminated. However, the Compact was going steady, and if anything, the state's lack of support strengthened the Compact (Murley 2018).

The new governor Ron DeSantis is expected to increase political support once more. Characterizing himself as a conservationist during the campaign (Miami Herald 2019a) has caused cautious optimism about his future endorsement of climate change mitigation as well as resilience. In his first week in office, Gov. DeSantis pledged US$2.5 billion to tackle problems like blue-green algae and red tide over the next four years. An executive order also established a state Office of Resiliency, a Blue-Green Algae Task Force, the Office of Environmental Accountability and Transparency and a new position called Chief Science Officer (Miami Herald 2019a). However, it remains to be seen how exactly DeSantis will support climate change mitigation and adaptation in the state.

On Miami-Dade County level, in 2015, groups such as The CLEO Institute and Miami Climate Alliance advocated for a higher allocation of resources for resilience within the County budget. At that time, the Office of Sustainability already existed but was staffed with three positions only. Mayor Carlos Gimenez then hired Jim Murley as the Head of the office and expanded the department to 11 positions to date. Miami-Dade County, together with the City of Miami Beach and the City of Miami, joined 100RC in 2016.

It is important to note that Miami-Dade County, unlike Broward County and Palm Beach County, has a strong Mayor, which means that the Mayor prepares and administers the city budget, and thus makes decisions on the allocation of funds to sustainability and resilience. Further, 1 million people currently live in MDC on unincorporated land and the County is providing municipal services to them. Thus, MDC arguably is in a stronger position to influence action on sustainability and resilience than the other Counties.

Also noteworthy for the following discussion is that the Counties can influence the use of the land in the county, including recommendations on zoning, subdivision, roads, uses of private property and areas and activities of state interest. Municipalities then decide on zoning in their city limits, for example permitting farmers markets or community agriculture projects in certain areas.

2.1.2 Immigration

A further determinant of Miami's diverse cultural, racial and sociodemographic fabric, which needs to be considered in the discussion of community resilience, is migration (both domestic and international). Recent US Census Bureau estimates show that the Sunshine State is now the third-most-populous state in the country, with 21.3 million residents, and it is growing (Census.gov 2019a). Migration from other states continues to be the prime contributor to population growth, but international migration plays an important role, particularly in the southern parts of the state. Florida has, by far, the largest net domestic migration in the nation and ranks third in international movements. In fact, migration has been responsible for 87% of Florida's population growth in this century. Because of the influx of retirees, its average age is much higher than elsewhere, and its births barely exceed deaths. In addition, Florida has a high number of "snow birds" (Northerners who spend their winters in the South) and other long-term vacationers.

Immigration has become a major contributor to population growth in Florida in recent years. Between 1980 and 1990, Florida had the nation's eighth fastest rate of increase in its foreign-born population—57%. Between 1985 and 1990 alone, some 350,000 entered the state legally. It is further estimated that each year from 1990 to 1995, Florida gained about 70,000 people through legal immigration. With continued large-scale illegal immigration, Florida's resident illegal population today could be as high as 450,000. Overall, the state has added 1.2 million people over the past five years, more than any other state except California and Texas, and at a higher rate of increase than those two larger states. One in five residents in the state is an immigrant, together making up more than a fourth of Florida's labor force (American Immigration Council 2018). The main countries where immigrants come from are Cuba, Haiti, Canada, Jamaica, Nicaragua, Colombia and the UK (Center for Immigration Studies 1995). Only 43% of Florida's immigrants have become US citizens.

Various parts of Florida are affected differently by immigration. The City of Miami (and the surrounding counties of Dade and Broward) represents the main magnet for immigrants. In 2017, its population was 473,000 and it is the 43rd largest city in the nation. But since 1970, little growth has occurred in the city itself; however, the Miami metropolitan area grew very rapidly.

Already in 1970, the city of Miami had a large foreign-born population and became one of the first cities in the nation to become a "minority-majority." Back then, 41.8% of Miamians were foreign born as the original immigration wave from Cuba had begun in the early 1960s. Immigration from Cuba has been particularly important for Miami because of the so-called wet foot, dry foot policy, which allowed anyone who emigrated from Cuba and entered the USA without a visa to pursue residency a year later. President Obama ended the policy in 2017 (CNN.com 2017).

The sources of immigration to Miami have shifted somewhat since 1970, though they have remained mainly Hispanic. During the 1980s and into the 1990s, large numbers of newcomers came from Nicaragua, Colombia, Peru and the Dominican Republic. Blacks came from Jamaica and Haiti. By 1990, there were 62% Hispanics, 12% Whites and 25% Blacks.

Of all of America's cities today, Miami is the largest city of the seven in the USA to have more immigrants than native born, and its foreign-born share (59.7%) is second only to Hialeah (70.4%). Hence demographically, Greater Miami is distinct. It is estimated that currently there reside people from over 100 countries, with about 60 languages spoken. These bring their culture and habits and thus create a rich diversity, which people in Miami are proud of (American Immigration Council 2018).

2.1.3 Community Development

Miami became an official city established on July 28, 1896 and started out with just 300 residents. The Florida East Coast Railway expanded into the agricultural terrains around that time and gave Miami its first economic advancement. By the 1920s, Miami was prospering well, but saw its first setback in 1925 with the real estate bubble burst. Before she could bounce back, the city was hit by the 1926 Miami Hurricane followed by the 1930s Great Depression. After aiding the battle against German submarines in World War II, Miami saw exceptional growth in its population which neared almost half a million. And it was under the control of Fidel Castro in Cuba around 1959 that Miami's population began to soar due to the migration of Cubans, as discussed earlier.

In order to understand the full spectrum of resilience challenges, particularly affecting Miami's diverse communities, one needs to consider the inequalities between neighborhoods and how they came about. One of the most notable events in this context, the consequence of which are still affecting Miami today, was the practice of "redlining."

In the twentieth century, the City of Miami imposed a "color line," limiting blacks residentially to a confined section of the city (Mohl 2001). Hence by the early 1930s, most of Miami-Dade's black population was situated in an area known at the time as "Colored Town," and today called Overtown, located just northwest of Miami's business district. The practice of "redlining" was triggered by the neighborhood appraisal and rating system developed by the Home Owners Loan Corporation (HOLC), established in 1933. And it led banks and other lending institutions to refuse mortgages and other loans to older, poorer, and minority neighborhoods, consequently leading to increased physical decay of these areas. Redlining was so consequential for Miami that Mohl (2001) notes that a 1987 map of Miami-Dade County showing projected black residential areas in 1990, matches almost exactly the 1938 HOLC map showing redlined areas of the county.

Throughout the 1930s, county and federal policies such as the New Deal housing policies seem to have exacerbated racial segregation in residential housing. Several studies conducted between 1956 and 1975 demonstrated that Miami-Dade County had the highest degree of residential segregation by race. The policies also aimed to remove blacks from city limits and especially from the high-value area in close proximity to the business district. Eventually, in 1946, a Florida circuit judge determined that Miami-Dade County did not have the power to enforce racial zoning, upholding the right of African Americans to purchase homes anywhere in the county, and the Dade County racial zoning ordinance was declared unconstitutional.

However, still in the 1950s, Dade County began implementing policies related to public housing and urban redevelopment, driven particularly by the construction of the interstate expressway 95 routed directly through Overtown and into downtown Miami, which forced African Americans from Overtown into new racially segregated neighborhoods such as Liberty City, Opa-locka, and Carol City. Completed in the mid-1960s, the I-95 downtown interchange alone eliminated the housing of about ten thousand people in Overtown. Only 8000 of 40,000 blacks who lived in Overtown before construction, now remain in the neighborhood. Mohl (2001) argues that well into the 1960s, the Miami Housing Authority (MHA), who controlled public housing in the county, used their authority with regard to site selection to maintain segregated housing patterns.

More recent development projects for upscale apartments and shopping centers resulted in the further destruction of public housing in Overtown. And by 1990s the northwest quadrant of the county had become primarily black. And gentrification continues to be a major challenge in Greater Miami, with and without being driven by climate change, as will be discussed below.

2.1.4 Food System

Greater Miami's food history was marked by immigration, consolidation, disconnected consumers and the proliferation of a digital revolution in food management.

Immigration, as discussed above, has influenced Greater Miami's food system on the production and consumption side. On the production side, the main agricultural area in Miami-Dade County is in Homestead. Many of the seasonal farmworkers stem from immigrant countries and are thus critical to sustain farming in the region. On the consumption side, local food preferences are being influenced by immigrant cultures and that has a significant impact on the uptake of healthful food options, as will be discussed below.

According to Charles LaPradd (2018), Miami-Dade County's Agricultural Manager, the food system in Miami-Dade has been consolidating on both the food buyer and supplier sides. On the buyer side, a consolidation of retail chains has led to the disappearance of a lot of independent retail stores. As a response, farmers started consolidating in order to keep up with the demand of the volume that buyers are asking for. Nowadays, some farmers buy their inputs and sell their produce through cooperatives or aggregators. Further, many farms went out of business, which has led to a further reduction in the number of farms. A similar development could be observed with ornamentals nurseries, the main agricultural activity in Palm Beach County.

At the same time, direct selling via Farmers Markets and U-Picks has increased by about 200% over the last decade, driven by consumers' demand for local and sustainable food, and out of a financial necessity of farmers. LaPradd estimates that today about 10% of total production in MDC is sold directly, through above-mentioned ways, as well as roadside stands.

Today, there is not only a problem relating to lack of access to healthy food, but also one of food commodification with consumers hardly knowing where their food is coming from, and usually not being interested.

On the food access side, in 2004 food stamps became electronic and paper coupons were replaced with electronic benefits and debit cards. Since Farmers' Markets had no wireless network, use of stamps dropped from US$9 million used to US$500,000. Although electronic benefits are now accepted at about one-third of markets in MDC and Broward, many low-income people still have not returned to visiting Farmers' Markets to use their SNAP benefits.

2.1.5 Climatic Events

Over the years, Greater Miami has experienced several environmental shocks and stresses it had to adapt to or bounce back from. Hurricane Andrew in 1992, another unnamed tropical storm in 1993, Hurricanes Wilma and Katrina in 2005 brought devastation and led to coastal erosion. King Tides caused major flooding after storms in 2011 and 2016. King Tides occur when the orbits and alignment of the Earth, moon, and sun combine to produce the greatest tidal effects of the year. King Tides bring unusually high water levels and can cause local tidal flooding. Over time, the rising sea level is increasing the height of tidal systems. Hence, most people in Greater Miami have lived through a severe weather event and thus believe that these events are increasing in number and severity.

Greater Miami is considered to be a "Ground Zero" location for rising sea levels. The Southeast Florida Regional Climate Change Compact estimates sea level rise of 6–10 inches by 2030, or 3–5 inches above average sea level in 2015. Predictions for the mid-term are between 11 and 22 inches of additional sea level rise by 2060, and longer-term between 28 and 57 inches, or up to 4.75 feet by 2100 (Southeastfloridaclimatecompact.org 2015).

Projections from Climate Central (2018) predict that with a 2-degree Celsius warming in global temperature and without any adaptation measures, by 2100 the South Florida peninsula could be flooded up to the city of West Palm Beach as illustrated in Fig. 2.1.

Miami's geology is a particularly complex obstacle in its battle against SLR and coastal erosion. Because the city's geology is composed of a porous, fossilized coral limestone, stormwater and high-tide events degrade the structural integrity of the roads, and the cities frequently experience extreme flooding as a result.

Just as natural systems worldwide are strained due to climate change, as sea levels continue to rise, beaches, mangroves and other man-made and natural systems in Greater Miami will likely be damaged from floods and storms. When eco-systems are damaged, their ability to protect the surrounding coastland is greatly impacted. The protections provided by offshore reefs and other marine systems are degrading due to ocean acidification and warming (Resilient305.org 2018). Moreover, the Everglades, which provide water to one-third of Floridians and irrigation for much of the state's agriculture (The National Wildlife Federation 2019), are threatened not only by an expansion of property development but by saline water intrusion, flooding, and consequent destruction.

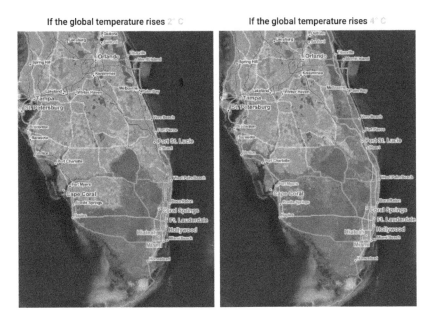

Fig. 2.1 How sea level rise could affect South Florida in 2100 (Climate Central 2018)

2.2 Climatic Stresses and Shocks

Below follows a discussion of the main stresses and shocks to Greater Miami's economy, environment, and its communities due to the warming climate.

2.2.1 Threats to Economy

The main industries driving GDP growth in Greater Miami are tourism, agriculture, and trade, and according to the Greater Miami Chamber of Commerce, currently the local economy is booming. Migration and immigration continue, as previously discussed, and hiring is expected to remain solid. Further, demand to Miami real estate continues to be strong (Miami Herald 2019b). However, it has been noted that if the impacts of climate change are left unchecked, there could be serious consequences for Florida's financial future (Southeastfloridaclimatecompact. org. 2017). As agriculture will be discussed in Sect. 2.4.2.1 as part of the food system, below potential impacts on Greater Miami's tourism, trade, and real estate industries will be discussed.

2.2.1.1 Tourism

As one of the state and county's most profitable industries, tourism revenues are essential to a healthy economy in Miami-Dade County, according to the Miami-Dade County Department of Regulatory and Economic Resources. The tourism industry employs circa 12% of all employees in Miami-Dade (The New Tropic 2015).

According to MiamiDade.gov, more people are coming to Miami than ever before. In 2017, Miami International Airport carried more than 40 million passengers, setting a new annual record. Add to that the 4.8 million multi-day travelers who took cruises leaving from Port Miami. Overall 14.6 million visitors stayed for at least one night in the Greater Miami area in 2014, spending money on lodging and shopping and visiting sights from Lincoln Road to the Everglades.

Most people who visit Miami do so for leisure and vacation purposes. Visitors from around the world come to enjoy the climate (the average temperature in South Florida is 76 degrees Fahrenheit) and Miami's beaches, experience the famed nightlife, and attend world-renowned events like Art Basel and Ultra Music Festival.

However, the effects of climate change threaten the tourism industry. As the majority of the region's tourist districts are located within low elevation zones, they are prone to flooding. More severe weather events can deter tourists in the future, and so can increased sunny day flooding and extreme heat.

2.2.1.2 Trade

Due to its low personal and corporate income tax level, its proximity to the Latin American market and good flight connections, the region is an attractive hub for international business. In fact, it ranks sixth in small business activity and number one in startup activity in the USA (Kauffman. org 2018). Half of the business in Miami-Dade is domestic (USA or Miami), half international. However, the region lacks a diversified economy (Resilient305.org 2018) with the main industry being real estate.

Just like other parts of the world, climate change will expose businesses in Greater Miami to higher risks related to both severe weather events such as hurricanes, plus the threat of increasingly often flooding. According to the Beacon Council, however, resilience is not yet a concern for businesses in Miami as the overall opinion is that MDC is preparing for the consequences of climate change. Further it is felt that Greater Miami as a business location may well be better off than other parts of the USA when it comes to the consequences of a warming climate such as severe weather events. As Jaap Donath, Ph.D., Senior Vice President Research and Strategic Planning at the Beacon Council described it during the interview for this book: *"Hurricanes can be foreseen days in advance, wildfires cannot."*

2.2.1.3 Real Estate
With its rapid urbanization and influx of international capital, Miami is one of the top real estate locations in the USA (Forbes.com 2018). Although land in Miami is considered such a valuable asset, according to Bloomberg (2017), SLR and hurricanes are beginning to affect the South Florida real estate market. Due to its low elevation and considering that there are, among the business developments in this location, 53,000 homes situated on land that is less than three feet above high tide, US$21 billion in assets are threatened by the rising sea levels.

Homes in Miami-Dade County near sea level are already seeing slower rise in value than those at higher elevations as buyers are waking up to the potential for frequent flooding, which will make it harder to sell in the coming years. And with the ongoing elevation of roads in some neighborhoods, residents are concerned that floodwaters will spill onto property that's not elevated, driving down the value of homes (Miami Herald 2018a).

While stronger building codes have helped protect newer buildings from storm surges, risks are costlier today due to higher water levels and increased development in vulnerable areas (Resilient305.org 2018). Storm and flood insurance rates are increasing. Latest with the introduction of new flood maps issued by the Federal Emergency Management Agency (FEMA) in 2020 (Miami Herald 2018b), and the expected subsequent rise of insurance premiums, a downward trend in new builds, and potentially sale in properties could be on the horizon soon.

The response of the real estate sector about the threats climate change poses is divided. Some local wealth managers already recommend to clients that they stop investing in property in Greater Miami (Wlrn.org 2018a). Other wealth managers and developers disagree. Even despite the recent report issued by the Intergovernmental Panel on Climate Change (IPCC), which predicts a climate catastrophe in the next couple of decades (Nytimes.com 2018), some developers still seek out opportunities by the water, or openly dismiss rising sea levels as paranoia (Bloomberg 2017).

2.2.2 Human Impacts

The effects of climate change on humans are harder to see than those on the economy. However, it already has affected people with chronic health problems. In Greater Miami, the hotter temperatures have caused people with breathing issues, and other ailments that are made worse by the heat, to limit their exposure to the sun by living in darkness in their own homes. Poorer patients often can't afford air conditioning or to visit cooler public places like libraries or malls (Wlrn.org 2018b).

Further, with the warming climate, there are new challenges related to the health of communities. A report from the Farmworker Association of Florida and the advocacy group Public Citizen shows that the temperatures in Florida were too hot at times in 2018 for workers who spend the majority of their time outdoors to do their work safely. The report shows that Florida has one of the nation's highest rates of heat-related hospital trips and more than half of 300 laborers (e.g., in construction or agriculture) surveyed didn't get any breaks in the shade (Miami Herald 2018f). Cheryl Holder, Program Director of FIU's Herbert Wertheim College of Medicine Panther Learning Communities (2018), notes increasing allergies, with the allergy season starting earlier and affecting particularly people with respiratory issues such as asthma. These can be found to higher rates in Miami's black communities. In addition, pollutants like ozone and the increase of the number of heat days exacerbate health issue. The heat brings pollutants closer to the ground, which increasingly exposes children to them. With no air conditioning in the Florida heat, lower-income residents open windows and doors and are more exposed than other parts of the population. Marc Mitchell, Associate Professor, Climate Change, Energy and Environmental Health Equity (2018) adds that according to a survey by the American Medical Association (AMA), 88% of doctors see the effects of climate change already. With more heat days, more people with preexisting conditions will die.

In addition, the warmer climate has led to a proliferation of diseases such as Zika, an illness caused by the Zika virus, which is spread by the Aedes mosquito. Infection with Zika during pregnancy can cause a birth defect called microcephaly and other severe fetal brain defects (CDC.gov 2019a).

2.3 Status Quo

To determine the current status of resilience in Greater Miami, one must look at both resource robustness and adaptive capacity. Hence in the following, this will be discussed both on the level of general resilience, as well as related to food and nutrition.

2.3.1 Community Resilience

2.3.1.1 Social Challenges
Greater Miami is facing considerable social challenges, which are being exacerbated by the stressors and shocks related to climate change. Among these present challenges is the discrepancy between the rich and poor, which according to Bloomberg is the highest in the USA (Bloomberg 2016). Though unemployment has been under 5%, it is estimated that 58% of Miami-Dade's population falls in the low-income range, struggling to pay for their basic needs due to the preponderance of lower-wage jobs and high cost of living (UnitedWayMiami.org 2017). Even the average income in Miami is 11% lower than the national average, with 20% of the population living below the poverty level—5% more than the national average (Resilient305.org 2018). And particularly the Youth is affected with 28% of people living in poverty are under the age of 18.

Further, 51% of households have insufficient liquid savings to survive for 3 months at the poverty level in the event of unexpected job or income loss. Job placement can be difficult for many residents battling barriers such as low education rates and skill sets needed for a changing economy (Resilient305.org 2018). Dade County schools are significantly overcrowded primarily because of continued immigration from Latin America. Only 12.8% of Miami adults have completed college, compared to the national average of 20.3% (American Immigration Council 2018).

A recent "The Color of Wealth in Miami" study that looked at the wealth gap between white households and households of color found large discrepancies between households based on color. Whereas white households have an estimated median net worth of US$107,000, for Black-American households, the median net worth was US$3700 (Insightcced.org 2019).

Figure 2.2 maps the density of residents living below the poverty line, as an illustration of particularly vulnerable neighborhoods.

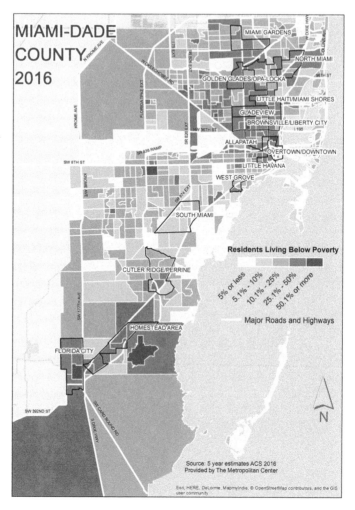

Fig. 2.2 Miami-Dade county 2014. Residents living below poverty (data analysis and mapping by FIU Metropolitan Center 2016)

For a population that is over 50% working poor and struggles to make ends meet, severe weather events, to proliferation of which is a result of a warming climate, are an additional stressor. In the event of storms, these communities cannot pack and evacuate like the more affluent parts of the population since, for instance they don't own a car, or cannot afford to pay for alternative accommodation or gas. Further, they can't buy weeks of supplies to weather the storm, and don't have financial resources to fix damages. Moreover, when children cannot go to school for three weeks since they are shut during the storm and recovery, parents can't afford to stay home, nor sometimes have the financial means to provide lunch at home which is otherwise provided at school.

2.3.1.2 Housing

Affordable housing is one of Miami's biggest resilience challenges. The local housing market is affected by external, often seasonal and absentee buyers. Approximately 29,000 new luxury residential units have been built since 2010.

At the same time, the cost burden for housing in Greater Miami has risen dramatically with average rents having increased 65% between 2009 and 2015. According to 2015 US Census estimates, the average Miami resident spends about 28% of their income on housing costs—a larger share than what residents pay for rent in any other city in the USA (Census.gov. 2019b).

In 2014, over 66% of renter households and over 50% of owner households were cost-burdened (meaning that they spend 30% or more on housing costs and struggle to pay other bills due to high housing expenses) making Miami the third least affordable housing market in the nation. Housing rights for tenants are lacking and the metropolitan area has one of the highest foreclosure rates in the nation. Homelessness has increased by 2% since 2015 (Resilient305.org 2018). Many people with lower incomes, especially in Miami's African-American and Hispanic communities, pay more than 50% of their income on rent (Census.gov. 2019b), making it a constant struggle for them to afford basics like nutritious food, health insurance, and transportation.

The Cities of Miami Beach and Miami are densely populated. Inland communities, such as Doral and Kendall, are experiencing increased mixed-use development. Although Miami-Dade County and the Cities of Miami, Hialeah, North Miami, Miami Beach, and Homestead receive funding from HUD for affordable housing. It is already with US$40 million far below the county's US$59 million cost to maintain housing for more than 30,000 residents and it is expected to be cut further (Miami Herald 2018e).

This affordable housing crisis becomes exacerbated by severe weather. A huge issue, for instance, is that the federal government does not require air conditioning (A/C) for public housing. With the increase of extreme heat days of 90 degrees and above, this becomes a public health hazard (Miami Herald 2018e). During Irma in 2017, which didn't even pass through Greater Miami, lives were lost due to power failures in critical A/C infrastructure in a retirement home in Hollywood. Since 2001, the county has required air conditioning in all redeveloped public housing projects. In the past five years, more than 2500 units have been rebuilt or have begun redevelopment, boosting the number of cooled homes across Miami-Dade (Miami Herald 2018e).

With the housing value decreasing along the South Florida coast as previously mentioned, climate gentrification will further displace the population. Climate gentrification involves an increase in real estate prices (and subsequently rents) in areas of higher elevations thus driving lower-income families into the more dangerous and less sustainable neighborhoods and exacerbating economic inequality (CNBC Markets 2018). Examples in Miami include Little Haiti, Overtown, and Allapattah. In Little Haiti, a neighborhood in the City of Miami, for example, longtime residents say they are getting calls and letters from prospective buyers who want to purchase their properties because of their elevation (Miami Herald 2018c).

The City of Miami Commission recently passed a resolution directing the administration to research gentrification in areas of low median income and high elevation, and ways to stabilize property taxes in these areas. The City will also dedicate US$4 million of the Miami Forever Bond to help residents at risk from climate change to make improvements on their homes. Miami is the first city in the USA to formally consider the topic (Miami Herald 2018c).

Another issue related to resilience of housing and livability in general is Miami's sewage system. In Greater Miami, two-thirds of residents live on underground septic tanks that could be failing by 2040 (Miami Herald 2019c). Rising sea and groundwater levels can affect onsite septic systems used to handle wastewater. Those located in low-lying areas face particular challenges, as septic system drain fields must be above the groundwater table and remain unsaturated to function effectively. Compromised septic systems have both public and environmental health risks, including potential contamination of potable water. Or as one interviewee put it: As sea levels continue to rise, toilets begin to fail and that's the end of civilization. A new study shows that rising seas have put tens of thousands of tanks at risk of failing within the next 20 years and that the most effective fix could cost more than US$3 billion (Miami Herald 2019c).

MDC is addressing this issue of potentially failing sewage systems by identifying current and future vulnerable areas and outlining potential approaches to limit vulnerability through infrastructure improvements and policy changes. Also, the City of Miami is updating its stormwater master plan to include recommendations for infrastructure improvements based on multiple sea level rise scenarios (Miamigov.com 2019). Overall, the water, sewer, and drainage systems need to be redesigned, rebuilt, and maintained to accommodate accelerating sea level rise and the potential for higher storm surges and heavy rainfall events (Resilient305.org 2018).

2.3.1.3 Mobility

Another challenge in Greater Miami that is particularly relevant to community nutrition resilience is mobility. Miami-Dade County is the sixth most congested county in the USA and connected, reliable transportation is needed between residential areas and jobs. Public transit is associated with several issues including safety, cleanliness, and reliability. In addition, and as opposed to other cities around the world, the public perceptions about transit need to change for more widespread ridership and dissolve the notion that transit is for the disenfranchised. Moreover, pedestrians and bicyclists feel unsafe competing with vehicles and transit ridership is declining in many areas. According to Resilient305.org (2018), more funding is needed for operations and maintenance of transportation infrastructure.

To address the transportation issue in the county, controversial expansions of main highways are planned, such as the expansion of SR 836 through the Everglades. Although providing a short-term solution to current traffic challenges, in the long term this project potentially poses a risk to the county's drinking water supply. Hence, transportation planning is an area that ultimately affects Greater Miami's communities' resilience.

2.3.2 Food Security

At present, the data for Greater Miami relating to food security at household level are fragmentary and inadequate. The "Map the Meal

Gap" report from Feeding South Florida (2017), as well as the ALICE report from UnitedWay (UnitedWayMiami.org 2017), and the Food Environment Atlas from the USDA Economic Research Services (2019) provide the best available data to draw conclusions on the status quo. Additionally, DataUSA.io (2019) and the US Department of Labor Bureau of Labor Statistics (2019) were used. The data for food security parameters, and food system access, availability and utilization are presented in Table 2.1.

Figure 2.3 illustrates a map of the neighborhoods in GM&B created by The Food Trust (2012). It displays lower-income communities with low supermarket sales with the goal of identifying the areas with the greatest need for more supermarkets. It shows that there is a direct correlation between poverty and food insecurity.

The findings presented above, as well as further research of the topic, allow for the following summary:

- The overall food insecurity rate is with 14.6 and 14.1% higher in Broward and Palm Beach than the National average of 12.6%. In Miami-Dade, it is lower. However, this doesn't tell the whole story as more individuals don't qualify for federal nutrition programs and therefore have to rely on emergency programs (Feeding South Florida 2018).
- Looking at SNAP participation as another proxy to food insecurity, it is with almost 18% double the official food insecurity rate. Studies have shown that individuals receiving SNAP benefits are significantly more likely to be food insecure than individuals not receiving SNAP benefits (Zeuli and Nijhuis 2017).
- Child food insecurity is higher in all three counties than the National average. One out of five children is food insecure; hence, their development opportunities are severely limited.
- Although the USDA ERS provides relatively low numbers related to the population in Miami-Dade and Broward that lives more than one mile from a supermarket (hence in a food desert), a Sun-Sentinel report from December 2015 called "The Hidden Hungry" found that 1.6 million people in South Florida were found to be living in 326 food deserts (Sun-Sentinel.com 2016).
- As can be expected, there is a direct correlation between poverty and food insecurity in the sense that low-income neighborhoods have less access to healthy food (or any kind of food for that

Table 2.1 Selected food security parameters for Miami-Dade, Broward, and Palm Beach Counties

Food security parameter	Miami-Dade	Broward	Palm Beach
Median HH income (US$/year)	$45,935	$54,212	$57,580
Food insecurity rate (%)	9	14.6	14.1
Child food insecurity rate (%)	20.6	19.1	20.5
Food purchase as $ of HH expenditure (%)	11.7		
% Limited access to healthy foods (%)	2	2	5
Population more than one mile from a supermarket or large grocery store (%)	6.51	11.23	20.47
Low-income population more than one mile from a supermarket or large grocery store (%)	1.81	2.75	4.88
Households without a car more than one mile from a supermarket or large grocery store (%)	0.35	0.57	0.85
Food system—access points			
Grocery stores per 1000 residents	0.24	0.2	0.19
Supercenters and club stores per 1000 residents	0.01	0.01	0.02
Convenience stores per 1000 residents	0.32	0.41	0.35
Specialized food stores per 1000 residents	0.08	0.07	0.07
SNAP-authorized stores per 1000 residents	0.6	0.57	0.54
Fast-food restaurants per 1000 residents	0.63	0.67	0.62
Full-service restaurants per 1000 residents	0.76	0.78	0.84
Farms with direct sales (%)	11.8	6.9	6.2
Direct farm sales (in $'000)	1913	77	954
Farmers' markets per 1000 residents	0.01	0.01	0.01
Farmers' markets that report accepting SNAP (%)	31	29	0
Farmers' markets that report selling fruit and vegetables (%)	50	71.43	50
Supplemental food offerings (data available on State-level only)			
SNAP participants (% of population)	17.73	17.73	17.73
National School Lunch Program participants (% of population)	8.34	8.34	8.34
School Breakfast Program participants (% of population)	3.90	3.90	3.90
Summer Food Service Program participants (% of population)	0.78	0.78	0.78

matter). The lack of full-service supermarkets in some communities means that residents must shop at convenience and corner stores with higher prices and more items of poor nutritional quality (The Food Trust 2012).

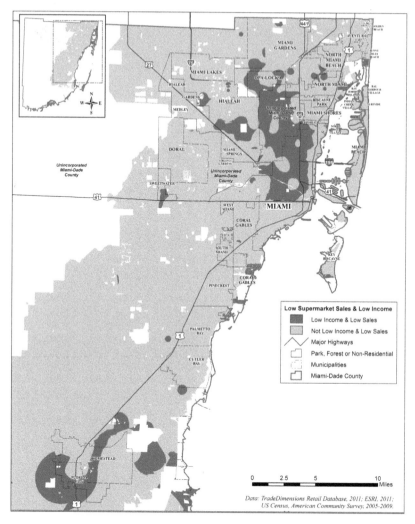

Fig. 2.3 Low supermarket sales and low income (The Food Trust 2012)

- There are considerable less farmers markets in all three counties than the National average of 0.04. In fact, Florida overall ranks 51 out of 50 US states plus Puerto Rico and D. C. on this parameter.

The Heifer's Locavore Index (Strollingtheheifers.com 2018), which compares states based on their commitment to healthy local food, ranks Florida 47 overall.
- There is with 50% also a surprisingly low percentage of farmers markets selling fruits and vegetables in Miami-Dade and in Palm Beach.
- School breakfast and summer food programs' participation is low compared to school lunch program participation, thus offering an opportunity for improvement in children's nutrition.

2.3.3 Health Implications

Nutrition security, or the lack thereof, directly impacts health parameters of Greater Miami's residents. Table 2.2 summarizes the main parameters related to health in the three Greater Miami counties. Data have been compiled from the Food Environment Atlas from the USDA Economic Research Services (2019).

In MDC, two-thirds of residents are overweight or obese (ABC Local 10 News 2014). Further, almost one in ten residents in the three counties has diabetes, type 2 of which is related to an unhealthy diet and lifestyle. Even more worrying is the high rate of obesity in low-income preschool children.

Studies demonstrate how better access to affordable nutritious food correlates to healthier eating (The Food Trust 2012):

- There is a 32% increase in consumption of fruits and vegetables among African Americans for every additional supermarket in a census tract.
- Adults with no supermarket within a mile of their homes are 25–46% less likely to have a healthy diet than those with the most supermarkets near their homes.
- Each additional meter of shelf space devoted to fresh vegetables is associated with an additional 0.35 servings of vegetables per day.
- Supermarket presence in a neighborhood is associated with decreased obesity and overweight, while convenience stores are associated with higher rates of diet-related disease.

Since Greater Miami's tax base must support this health epidemic, it makes sense to look at the economic case in more detail. For example, when looking at one theoretical resident Mrs. Jones. Mrs. Jones is

Table 2.2 Selected health parameters for Miami-Dade, Broward, and Palm Beach Counties

Health parameter	Miami-Dade (%)	Broward (%)	Palm Beach (%)
Adult obesity rate, 2013	26.4	26.4	26.4
Adult diabetes rate, 2010	8.8	9.2	9.5
Low-income preschool obesity rate, 2009–2011	16.1	11	15.6

42 years old and has type 2 diabetes primarily caused by an unhealthy diet and little exercise. She further has osteoporosis and heart issues that have kept her in hospital for a few nights last year. Hence, her bad diet is costing taxpayers around US$850,000 per year, as her health-related expenses are paid through the public-assistance program Medicaid.

Extrapolating her case to one neighborhood, let's say a block in Allapattah, there could be twenty residents with similar health issues, hence costing around US$16 million. These cost estimates can now be considered when structuring interventions and looking at return on investment of these. For example, creating a program for healthy eating for the community for about US$40,000 that reduce health costs in the community, even by only a fraction, seems like a worthwhile investment and clearly demonstrates the importance of taking action on nutrition resilience.

2.4 Threats to Community Nutrition Resilience in Miami

Challenges to Miami's community nutrition resilience can be found primarily associated with the lack of focus on food and nutrition in the region, systemic stresses in the food system as well as shocks due to climate change.

2.4.1 Lack of Focus, Policy, and Planning

Overall, the focus of local officials and other critical parties in Greater Miami on food security and community nutrition resilience can be described as minimal, if not nonexistent. These are discussed below

For example, there is currently no food policy council in South Florida. Food policy councils are designed to influence local and state food policy and typically include representatives from across the food system (Zeuli and Nijhuis 2017). There are approximately 200 food policy councils in the USA. The City of Miami had a food policy council in the past, which managed to get a permit for farmers markets, which was later adopted by the county. The council however dissolved due to conflicting agendas of participants and a perceived lack of ability to influence policy. However, a food policy council can help decrease some of the difficulties that the current local policies are posing on food business, for example.

Also related to inadequate food policy is the lack of support for local food businesses. In an interview with Della Heiman, the creator of Wynwood Yard (the Yard)—a pioneering foodpreneur incubator and local food truck space, she explained her challenges when dealing with the permitting process to set up the Yard. The Yard employees 150 people mostly from the local neighborhood Wynwood and has over 40 businesses incubated, yet there has been close to no support from the city to set up or continue the Yard.

According to Heiman, the City of Miami's permitting preceding construction was extremely difficult, requiring her to visit the Zoning department almost every single day and included escalating her case to the department heads as staff was not familiar with the circumstances of her application. She also recounted having been to the City Commission at least 20 times.

In addition, once the Yard was set up and running, it had to deal with almost daily visits from city inspectors, sometimes at 3 o'clock in the morning. And although they were never fined for any breach, dealing with these interruptions to daily operations each day was challenging for the new business.

Equally during the Zika outbreak in Miami, there was no support to minimize the loss of business, neither was there any during nor after the destruction caused by storms Irma in 2017 and Matthew in 2016. In fact, Heiman recalls that after Irma, the Yard was one of the last establishments in her direct neighborhood to be reconnected to electricity. Overall, these high barriers posed by local government, coupled with lack of support, and extremely high costs of starting a food business in Miami, make it extremely difficult for local, innovative, and culturally relevant foodpreneurs to enter and stay in the market.

Equally, Art Friedrich from Urban Oasis Project, a local nonprofit dedicated to promoting healthy and local food, particularly through community food gardens and farmers markets, echoes that there has been no support for complimentary food access points such as farmers markets from elected officials.

According to Friedrich, farmers markets are clearly not a priority for the cities in MDC, and organizations in the space have done a lot of education about the benefits for communities. He echoes the challenges particularly related to dealing with the zoning department.

Furthermore, urban planning in MDC currently does not include food systems and according to a representative from the MDC Planning and Zoning Department is not expected to in the near future.

In addition to the lack of focus from officials, there is very little focus by Foundations, which traditionally have been strong in working on challenges in Greater Miami. The only notable exception is the Health Foundation of South Florida which supports the Live Healthy Miami Gardens and Live Healthy Little Havana initiatives.

2.4.2 Threats Related to the Food System

At this point, an analysis of food systems in the three Greater Miami counties (as illustrated in Fig. 1.1) and its vulnerabilities is needed. However, there has been no such analysis conducted so far. Hence, this book will attempt to cover the constituent parts of the system from food production, via food processing, food distribution, to food access, and recovery. It will focus primarily on Miami-Dade County as a starting point.

2.4.2.1 Food Production

Florida's agriculture is important for the food security of the entire nation. With its unique soil structure, climate, and crops that can be grown, and a 12-month growing cycle, agriculture is a major asset that needs to be protected. Even though currently it may be not financially viable to grow certain crops, they could be converted if supply from other sources were ever interrupted. Hence, maintaining Florida's agriculture strong increases food security and resilience for the country.

With about US$600–US$700 million a year, the value of crops grown in Miami-Dade (mainly traditional vegetables) is number two in the State of Florida and number 22 nationwide. Agriculture is also the number 2 industry in MDC. Palm Beach County's agriculture industry, with US$1 billion in sales of crops at the farm level and a US$10 billion economic impact, ranks first in Florida and makes up more than a third of the county's 2000 square miles (The Palm Beach Post 2016). For instance, for wintertime vegetables there are only two places in the USA where can you produce these products, Southern California and Southern Florida. However, according to LaPradd (2018), recently production particularly of wintertime vegetables has scaled down because of competition with Mexico.

Since the USA entered into a trade agreement with Mexico in 1992, there has been a steady decline in market demand for traditional vegetables from South Florida. Where in the past, Miami-Dade growers farmed about 50–60% vegetables, with fruits and ornamentals accounting for the rest, now ornamentals account for 50–60%, and fruits and vegetables for the rest.

According to LaPradd (2018), the agricultural industry in Miami-Dade is already feeling the effects of climate change. And climate disruptions to agriculture in Florida are projected to increase in the future. These impacts will be increasingly negative because critical thresholds are being exceeded (Letson 2017). In the coming decades, it is expected that agriculture will be affected by invasive alien species, sea level rise flooding, and storm surges. Already now a higher rate of insect and fungus infestations can be observed, one reason being that it doesn't get cold enough anymore to interrupt the lifecycle of major pests such as whiteflies (LaPradd 2018).

A warmer, drier climate will also place agriculture in competition with other users for limited water resources. Equally, seasonal agriculture will continue to require careful water management, especially given saltwater intrusion into underground aquifers (Resilient305.org 2018). Further the degradation of soil and water assets due to increasing extremes in precipitation will challenge both rainfed and irrigated agriculture. High night-time temperatures can also reduce grain yields and animal-sourced production.

Other challenges to agriculture in Miami-Dade that have been mentioned in interviews are:

- The use of arable land for solar farms, in order to meet new renewable energy targets;
- The challenge to attract talent to the industry;
- Issue of land ownership;
- The high interest of farmers to sell their land profitably to developers;

- The need to increase agricultural innovation to adapt to new climate reality; and
- A fragmented industry of individualistic farmers that do not like interference and/or are not open to external support.

Aquaculture in the county is also at risk, as salmon, for example, is being farmed with underground water. This threatens companies such as Atlantic Sapphire USA, a subsidiary of Norwegian farmed salmon firm Atlantic Sapphire A/S, has built a US$350 million land-based aquaculture facility in Miami (Seafoodsource 2017). Equally, subsistence fishing is challenged as the canals get polluted. And a number of communities rely on subsistence fishing.

2.4.2.2 Food Processing

On the processing side, there are no major food processing facilities in Greater Miami that could be under threat of SLR.

According to LaPradd (2018), following the market consolidation discussed in Sect. 2.1.4, now fresh produce is bought by only a handful of buyers. Consequently, small growers cannot deliver their crops independently to retail but must go through a packing house/wholesale. Equally, terminal markets have disappeared and there has been a move away from independent buyers. This decreases overall resilience due to a lack of redundancy and diversity.

2.4.2.3 Food Distribution

Resilience challenges related to food distribution include market barriers for local farmers looking to offer their fresh produce directly to local retailers. The main retailers in Greater Miami are Walmart, Trader Joe's, Presidente, Aldi, Whole Foods, Publix, Winn Dixie, and Milam's Market.

According to Art Friedrich, President of Urban Oasis Project (2018), about 25 years ago Publix would frequently purchase from local growers. Today, their purchasing process is extremely complex for farmers to work with; hence, very little to no produce is purchased by the retailer directly from farmers.

Whole Foods appears to be more flexible when it comes to working with growers directly, although they don't buy much produce locally. A barrier for local farmers is that they need constant volumes to supply to any retailer, and that's close to impossible for a single farm to comply with.

Further, the costs of suppliers from abroad are lower. Internet sales, e.g., through Amazon, are interesting for farmers but difficult to operate, according to LaPradd (2018). Hence, it appears that resilience due to diverse routes of selling has been reducing for farmers. One can assume that this also leads to lower nutrition resilience for consumers since they mainly rely on supply from only a few large retailers.

2.4.2.4 Food Access

Access-related challenges to community nutrition resilience include the availability of, and affordability in food access points. An issue directly related to food distribution is unequal pricing of retail stores in different neighborhoods. Several studies (e.g., Gosliner et al. 2018; MacDonald and Nelson 1991) have found that fresh produce is more expensive in low-income neighborhoods and that convenience stores offer more expensive, poorer-quality produce than other stores. That is mainly due to a lack of competition in lower-income neighborhoods and suggests that policy makers should consider incentivizing retail chains to open stores in these communities.

In terms of complimentary access points, farmers markets (stationary and mobile), community-supported agriculture (CSA) and U-Picks can be mentioned. Farmers markets are examples of retail mechanisms that support direct sales of fresh produce from local farms to community members. According to the US Department of Agriculture (USDA), the total number of registered farmer markets increased by 35% between 2009 and 2013. Singleton et al. (2015) found that disparities in the availability of farmers markets exist. For example, there are fewer markets in areas with higher non-Hispanic black residents and residents living in poverty. Also, the more fast-food restaurants in an area, the less markets. According to the USDA Economic Research Services (2019), in Miami-Dade there were 32 farmers markets in 2013. In Broward, there were 14, and in Palm Beach, there were 10.

CSAs are another option for consumers to access locally grown fresh produce for a lower price. In the case of CSAs, participants pay in advance before the start of the season for a share of the farm's harvest. In turn, they receive a portion of farm products each week (or biweekly) throughout the growing season. Prepayment gives the small farmer operating capital for the year's production and helps the farmer plan the planting and harvest schedule. Shareholders not only share the bounty from the farm, they also share the risk. If there is a crop failure (such as from disease, a hurricane or major freeze), the shareholders agree to take the loss with the farmer. In South Florida, the season lasts about 5 months from late November through mid-April (Beeheavenfarm.com 2018). According to LaPradd, recently there has been a decrease in sign-ups for CSA in Miami-Dade as there have been cases of quality issues. Further, meal delivery services have taken some of the shares.

U-picks, farm stands, or roadside stands are another way of direct sales of local farmers to consumers, and often a more profitable one for farmers as they are not incurring distributor margins. For consumers, these options add an experience of interaction with the farmer as produce is picked up on, or close to, the farm. Prominent examples are Knaus Berry Farm, or Robert Is Here in Homestead.

2.4.2.5 Food Recovery

Up to 40% of edible food produced in the USA ends up in landfills every year. In Florida, and without harvest losses, over 700 million pounds of produce goes to waste every year. This waste produces GHG that contribute to climate change. In addition, wasting food means wasting the resources such as soil and water that have gone into its production. Meanwhile, as discussed earlier, over 48 million people in the USA are food insecure. In the aftermath of a disaster, households that are already food insecure face additional challenges, while others may become food insecure due to disaster-related expenses and hardships. Food recovery and redistribution could play an interesting role in alleviating some of the systemic and event-related insecurities.

2.4.2.6 Cultural Relevance

Looking at cultural relevance of food is important for the uptake of the offered options. For many immigrant cultures in Miami vegetables are not part of their culture. Many don't know the vegetables offered in the region and consequently don't know how to make them taste good. LaPradd (2018) seconds that there are groups of immigrants in MDC that are used to diets with very little fresh produce. Hence, even if fresh produce is offered, it is not necessarily used by consumers. And this means the challenge of offering culturally relevant healthy food is important for a community nutrition resilience strategy.

Another argument is that the original lifestyle of immigrant families is not what would make them obese and ultimately sick. Thi Squire, the Executive Director of Grow2Heal a community nutrition program by

Baptist Health supports this notion that it is in fact the adoption of a Western diet, as well as convenience factors such as fast-food restaurants. A study by University of Miami (ABC Local 10 News 2014) supports this as it found that particularly Cubans, who would walk a lot in their home country come to Miami and depend on alternative transport and consequently gain weight. Both viewpoints should be considered when designing programs that address community nutrition resilience in a culturally relevant way.

2.4.3 Health Literacy and Education

Related to cultural relevance is health literacy. Health literacy can contribute significantly to making better food choices. The Patient Protection and Affordable Care Act of 2010, Title V, defines health literacy as the degree to which an individual has the capacity to obtain, communicate, process, and understand basic health information and services to make appropriate health decisions (Cdc.gov 2019b). Related to nutrition and in simple terms, it's the ability of people to make the connection between the food they eat and their health. Updated numbers on health literacy in Greater Miami could not be found. However, the 2003 National Assessment of Adult Literacy (NAAL) suggests a US-wide health literacy rate of only 12% (Health.gov 2019). Hence, effective promotion and messaging campaigns to increase that percentage in Greater Miami are needed.

And even where education is offered, it is often misleading. An example is MyPlate, promoted by the US Department of Agriculture as the recommended healthy eating style. Despite being an improvement from the previously advocated MyPyramid, it has been criticized for not offering the most complete picture when it comes to basic nutrition advice (Harvard TH Chan School of Public Health 2019) mainly because it still appears heavily influenced by political and commercial pressure from industry lobbyists. However, it still is used by many doctors and publically run institutions as the guiding principles for a healthy and balanced nutrition, hence providing a distorted type of health literacy to residents of Greater Miami.

Equally, the marketing of food companies can be misleading. It exacerbates the issue as it often focuses on convenience and price, and thus not working on literacy but motivates people to make unhealthy food choices and lose their cooking skills.

Locally, the Institute for Food and Agricultural Standards of University of Florida (IFAS) addresses the issue of health literacy by educating community groups, school staff, and lawmakers on healthy nutrition.

According to Squire (2018), the main challenges in adopting a nutritious diet are the lack of cooking skills driven by a quest for convenience. Telling people what to eat does not make them change their habits. Instead understanding their culture and why they eat what they eat, even if it makes them sick, is a big part of the solutions. Subsequently, programs need to focus on teaching them how to cook and enjoy healthy food (Squire 2018).

Equally, people are drawn to the food they grow up with. James Jiler (2018), who worked with inmates in Rikers Island for 10 years, reports that inmates that were released were craving a McDonald's Big Mac, and that was more important to them than meeting with their family or partner, for example. Hence, healthy food needs to be introduced at an early age.

In addition to improving health outcomes in the long run, education about food, its origin, and properties, especially about local fresh produce, can help address the process of decommodification of nourishment, with people caring less about where in the world the food they eat came from, who produced it and how (Wilson 2012).

2.4.4 Adaptive Capacity

Following Longstaff et al. (2010)'s conceptualization of community resilience, resilient communities must have both the resources available and the ability to apply or reorganize them in such a way to ensure essential functionality. Hence in addition to providing communities with resources such as food access points and education about health, they need to be enabled and empowered to use them. This includes creating networks within the community, so residents can help each other. Further, it includes knowledge about how to advocate for community needs with local governments and access to financial resources for community-led programs.

When natural disasters strike, the resilience particularly of lower-income neighborhoods collapses. After hurricane Irma in 2017, Miami's poorest neighborhoods went days without electricity and subsequently were not able to access food because benefit cards from the SNAP can only be used with functioning credit card systems. Simultaneously, restaurants and supermarkets had to throw out large amounts of food, which could have been used to feed the ones in need of food. In some instances, this happened as barbecues were organized in some underprivileged neighborhoods with fresh meat and seafood that Miami Beach restaurants were no longer allowed to sell (Wlrn.org 2017). This example indicates that one important avenue to evaluate as a solution to food resilience challenges in Miami is where local small businesses can offer assistance during post-disaster recovery periods. When evaluating the potential for business owners to take on this responsibility, however, one should take into consideration to what extent this exposes their vulnerabilities and what measures can and should be taken to limit such concerns.

Because of all of the challenges discussed above, we need to look at ways to improve systemic as well as event-related nutrition resilience in Greater Miami. Chapter 3 will now review the actions Greater Miami is taking to address systemic stressors to its community nutrition resilience, as well as prepare for and adapt to sudden shocks. Next, Chapter 4 will offer an analysis of a few case studies of other urban regions that can inform actions on community nutrition resilience going forward. Finally, Chapter 5 summarizes this book and gives tangible recommendations to practitioners in Greater Miami, and beyond, on how to strengthen community nutrition resilience.

REFERENCES

100ResilientCities.org. (2018). *Frequently asked questions (FAQ) about 100 resilient cities* [online]. Available at: http://www.100resilientcities.org/100RC-FAQ/#/-_/. Accessed 29 January 2018.

ABC Local 10 News. (2014). *Report: 2/3 of Miami-Dade County residents overweight, obese* [online]. Available at: https://www.local10.com/news/florida/miami-dade/report-2_3-of-miami-dade-county-residents-overweight-obese. Accessed 21 December 2018.

American Immigration Council. (2018). *Immigrants in Florida* [online]. Available at: https://www.americanimmigrationcouncil.org/research/immigrants-florida. Accessed 10 December 2018.

Beeheavenfarm.com. (2018). *CSA: Info* [online]. Available at: http://beeheavenfarm.com/csa/info/. Accessed 19 October 2018.

Bloomberg. (2016). *Benchmark: The 10 most unequal cities in America: The South Florida city is neck and neck with Atlanta and New Orleans* [online]. Available at: https://www.bloomberg.com/news/articles/2016-10-05/miami-is-the-newly-crowned-most-unequal-city-in-the-u-s. Accessed 13 December 2018.

Bloomberg. (2017). *South Florida's real estate reckoning could be closer than you think* [online]. Available at: https://www.bloomberg.com/news/features/2017-12-29/south-florida-s-real-estate-reckoning-could-be-closer-than-you-think. Accessed 29 January 2018.

Carpenter, S. R., Westley, F., & Turner, G. (2005). Surrogates for resilience of social-ecological systems. *Ecosystems, 8*(8), 941–944. https://doi.org/10.1007/s10021-005-0170-y.

Cdc.gov. (2019a). *Zika virus: Pregnancy* [online]. Available at: https://www.cdc.gov/zika/pregnancy/index.html. Accessed 13 February 2019.

Cdc.gov. (2019b) *What is health literacy* [online]. Available at: https://www.cdc.gov/healthliteracy/learn/index.html. Accessed 2 February 2019.

Census.gov. (2019a). *Quick facts Florida* [online]. Available at: https://www.census.gov/quickfacts/fl. Accessed 2 February 2019.

Census.gov. (2019b). *American housing survey—Table creator* [online]. Available at: https://www.census.gov/programs-surveys/ahs/data/interactive/ahstablecreator.html. Accessed 4 February 2019.

Center for Immigration Studies. (1995). *Shaping Florida: The effects of immigration, 1970–2020* [online]. Available at: https://cis.org/Report/Shaping-Florida-Effects-Immigration-19702020. Accessed 10 December 2018.

CNBC Markets. (2018). *Rising risks: 'Climate gentrification' is changing Miami real estate values—For better and worse* [online]. Available at: https://www.cnbc.com/2018/08/29/climate-gentrification-is-changing-miami-real-estate-values.html. Accessed 14 September 2018.

CNN.com. (2017). *US ending 'wet foot, dry foot' policy for Cubans* [online]. Available at: https://www.cnn.com/2017/01/12/politics/us-to-end-wet-foot-dry-foot-policy-for-cubans/index.html. Accessed 2 February 2019.

DataUSA.io. (2019). *Find a profile* [online]. Available at: https://datausa.io/search/?kind=geo. Accessed 10 January 2019.

Feeding South Florida. (2017). *Map the meal gap 2017: A landmark analysis of food insecurity in the United States* [online]. Available at: http://dev.feedingsouthflorida.org/wp-content/uploads/2017/05/Map-the-Meal-Gap-Feeding-South-Florida-2017.pdf. Accessed 5 February 2019.

Feeding South Florida. (2018). *South Florida continues to face hunger challenges* [online]. Available at: https://feedingsouthflorida.org/south-florida-continues-to-face-hunger-challenges/. Accessed 20 February 2019.

Forbes.com. (2018). *15 booming real estate markets that are trending in 2018* [online]. Available at: https://www.forbes.com/sites/forbesrealestatecouncil/2018/04/03/15-booming-real-estate-markets-that-are-trending-in-2018/#2db0add226f6. Accessed 20 December 2019.

Gosliner, W., Brown, D., Sun, B. C., Woodward-Lopez, G., & Crawford, P. B. (2018). Availability, quality and price of produce in low-income neighbourhood food stores in California raise equity issues. *Public Health Nutrition, 21*(9), 1639–1648. https://doi.org/10.1017/S1368980018000058.

Harvard TH Chan School of Public Health. (2019). *Healthy eating plate vs. USDA's MyPlate* [online]. Available at: https://www.hsph.harvard.edu/nutritionsource/healthy-eating-plate-vs-usda-myplate/. Accessed 13 February 2019.

Health.gov. (2019). *Quick guide to health literacy* [online]. Available at: https://health.gov/communication/literacy/quickguide/factsbasic.htm#six. Accessed 23 May 2019.

Insightcced.org. (2019). *The Color of Wealth in Miami* [online]. Available at: https://insightcced.org/the-color-of-wealth-in-miami/. Accessed 2 March 2019.

Kauffman.org. (2018). *The Kauffman index: Miami-Fort Lauderdale-Pompano Beach* [online]. Available at: https://www.kauffman.org/kauffman-index/profile?loc=33100&name=miami-fort-lauderdale-pompano-beach&breakdowns=growth|overall,startup-activity|overall,main-street|overall. Accessed 20 December 2018.

LaPradd, C. (2018, November 2). Personal interview.

Letson, D. (2017). Climate change and food security: Florida's agriculture in the coming decades. In A. Schmitz, P. L. Kennedy, & T. G. Schmitz (Eds.), *World agricultural resources and food security* (Frontiers of Economics and Globalization, Vol. 17, pp. 85–102). Bingley: Emerald Publishing. https://doi.org/10.1108/s1574-871520170000017007.

Longstaff, P. H., Armstrong, N. J., Perrin, K., Parker, W. M., & Hidek, M. A. (2010). Building resilient communities: A preliminary framework for assessment. *Homeland Security Affairs, VI*(3), 1–23.

MacDonald, J. M., & Nelson, P. E. (1991). Do the poor still pay more? Food price variations in large metropolitan areas. *Journal of Urban Economics, 30*, 344–359.

Miamigov.com. (2019). *Coastal and stormwater infrastructure* [online]. Available at: https://www.miamigov.com/Government/MiamiClimateSolutions/Coastal-and-Stormwater-Infrastructure. Accessed 2 March 2019.

Miami Herald. (2018a). *'Why are we the guinea pig?': Climate change project divides a Miami Beach neighborhood* [online]. Available at: https://www.miamiherald.com/article223054220.html. Accessed 2 December 2018.

Miami Herald. (2018b). *Your flood insurance premium is going up again, and that's only the beginning* [online]. Available at: https://www.miamiherald.com/news/state/florida/article215162440.html. Accessed 4 September 2018.

Miami Herald. (2018c). *Climate gentrification: Is sea rise turning Miami high ground into a hot commodity?* [online]. Available at: https://www.miamiherald.com/news/local/environment/article222547640.html. Accessed 3 January 2019.

Miami Herald. (2018d). *Kids are suing Gov. Rick Scott to force Florida to take action on climate change* [online]. Available at: https://www.miamiherald.com/news/local/environment/article208967284.html. Accessed 10 December 2018.

Miami Herald. (2018e). *It's really hot in Miami, but the feds don't require A/C in public housing* [online]. Available at: https://www.miamiherald.com/news/local/community/miami-dade/edison-liberty-city/article217030505.html. Accessed 30 October 2018.

Miami Herald. (2018f). *Florida heat is already hard on outdoor workers: Climate change will raise health risks* [online]. Available at: https://www.miamiherald.com/news/local/environment/article220833555.html. Accessed 20 December 2018.

Miami Herald. (2019a). *DeSantis announces sweeping fixes meant to clean up Florida water woes* [online]. Available at: https://www.miamiherald.com/news/local/environment/article224219365.html. Accessed 30 January 2019.

Miami Herald. (2019b). *Threat to Miami economy from stock trouble, shutdown— And robots—Is limited, experts say* [online]. Available at: https://www.miamiherald.com/news/business/article224623860.html. Accessed 10 February 2019.

Miami Herald. (2019c). *A $3 billion problem: Miami-Dade's septic tanks are already failing due to sea rise* [online]. Available at: https://www.miamiherald.com/news/local/environment/article224132115.html. Accessed 15 January 2019.

Mohl, R. A. (2001). Whitening Miami: Race, housing, and government policy in twentieth-century Dade County. In *The Florida historical quarterly, vol 79, no. 3: In reconsidering race relations in early twentieth-century Florida (Winter)* (pp 319–345). Florida: Historical Society.

NCA2018.Globalchange.gov. (2019). *Fourth national climate assessment* [online]. Available at: https://nca2018.globalchange.gov. Accessed 1 February 2019.

Nytimes.com. (2018). *Major climate report describes a strong risk of crisis as early as 2040* [online]. Available at: https://www.nytimes.com/2018/10/07/climate/ipcc-climate-report-2040.html. Accessed 20 December 2018.

Resilience Alliance. (2007). *Assessing and managing resilience in social-ecological systems: A practitioner's workbook (resilience alliance)*. Available at: www.resalliance.org.

Resilient305.org. (2018). *Resilient Greater Miami & the beaches: Preliminary resilience assessment #Resilient305* [online]. Available at: http://resilient305.com/assets/pdf/170905_GM&B%20PRA_v01-2.pdf. Accessed 29 January 2018.

Seafoodsource. (2017). *Atlantic Sapphire building USD 350 million land-based salmon farm in Miami* [online]. Available at: https://www.seafoodsource.com/news/aquaculture/atlantic-sapphire-building-usd-350-million-land-based-salmon-farm-in-miami. Accessed 2 March 2019.

Singleton, C. R., Sen, B., & Affuso, O. (2015). Disparities in the availability of farmers markets in the United States. *Environmental Justice (Print), 8*(4), 135–143. https://doi.org/10.1089/env.2015.0011.

Southeastfloridaclimatecompact.org. (2015). *Unified sea level rise projection. Southeast Florida* [online]. Available at: http://www.southeastfloridaclimatecompact.org/wp-content/uploads/2015/10/2015-Compact-Unified-Sea-Level-Rise-Projection.pdf. Accessed 2 March 2019.

Southeastfloridaclimatecompact.org. (2017). *Advancing resilience solutions through regional action* [online]. Available at: http://www.southeastfloridaclimatecompact.org/. Accessed 18 December 2017.

Squire, T. (2018, November 19). Personal interview.

Strollingtheheifers.com. (2018). *Locavore index: How locavore is your state?* [online]. Available at: https://www.strollingoftheheifers.com/locavore/. Accessed 20 February 2019.

Sun-Sentinel.com. (2016). *Florida lawmakers consider low-cost grocery store loans to promote healthy eating* [online]. Available at: http://www.sun-sentinel.com/health/fl-food-desert-supermarket-legislation-20160205-story.html. Accessed 29 January 2018.

The Food Trust. (2012). *A healthier future for Miami-Dade County: Expanding supermarket access in areas of need* [online]. Available at: http://thefoodtrust.org/uploads/media_items/miami-dade-supermarket-report.original.pdf. Accessed 26 January 2019.

The National Wildlife Federation. (2019). *The Everglades* [online]. Available at: https://www.nwf.org/Educational-Resources/Wildlife-Guide/Wild-Places/Everglades. Accessed 2 March 2019.

The New Tropic. (2015). *How tourism impacts Miami* [online]. Available at: https://thenewtropic.com/tourism-economy-culture/. Accessed 10 December 2018.

The Palm Beach Post. (2016). *Palm Beach County agriculture touted as economic engine at summit* [online]. Available at: https://www.palmbeachpost.com/article/20160504/BUSINESS/812068300. Accessed 2 March 2019.

United States Department of Labor Bureau of Labor Statistics. (2019). *Southeast Information Office* [online]. Available at: https://www.bls.gov/regions/southeast/news-release/consumerexpenditures_miami.htm. Accessed 20 January 2019.

UnitedWayMiami.org. (2017). *The 2017 united way ALICE report* [online]. Available at: https://unitedwaymiami.org/wp-content/uploads/2014/11/10351-EXT-ALICE-2017-FINAL-single-pgs-1.pdf. Accessed 4 January 2019.

USDA Economic Research Service. (2019). *Food environment Atlas* [online]. Available at: https://www.ers.usda.gov/data-products/food-environment-atlas/. Accessed 5 February 2019.

Whitehouse.gov. (2018). *OMB bulletin no. 18-03* [online]. Available at: https://www.whitehouse.gov/wp-content/uploads/2018/04/OMB-BULLETIN-NO.-18-03-Final.pdf. Accessed 2 February 2019.

Wilson, A. D. (2012). Beyond alternative: Exploring the potential for autonomous food spaces. *Antipode, 45*(3), 719–737. https://doi.org/10.1111/j.1467-8330.2012.01020.x.

Wlrn.org. (2017). *After hurricane Irma, food insecurity in Miami-Dade's poorest communities* [online]. Available at: http://wlrn.org/post/after-hurricane-irma-food-insecurity-miami-dades-poorest-communities. Accessed 29 January 2018.

Wlrn.org. (2018a). *Monday was a big day in climate and economic news. Here's what South Floridians should know* [online]. Available at: http://www.wlrn.org/post/monday-was-big-day-climate-and-economic-news-heres-what-south-floridians-should-know. Accessed 10 February 2019.

Wlrn.org. (2018b). *Temperatures in Florida are rising: For vulnerable patients, the heat can be life-threatening* [online]. Available at: http://www.wlrn.org/post/temperatures-florida-are-rising-vulnerable-patients-heat-can-be-life-threatening. Accessed 20 December 2018.

Zeuli, K., & Nijhuis, A. (2017). *The resilience of America's urban food systems: Evidence from five cities* [ebook]. Roxbury, MA: ICIC. Available at: http://icic.org/wp-content/uploads/2017/01/Rockefeller_ResilientFoodSystems_FINAL_post.pdf?x96880. Accessed 18 December 2017.

CHAPTER 3

Taking (Community Nutrition) Resilience Action

Abstract This chapter examines how Greater Miami is responding to the resilience challenges introduced in Chapter 2. It looks at political actions on federal, state, and county levels and discusses three key initiatives guiding resilience action in the region: the Southeast Florida Regional Climate Change Compact, Rockefeller Foundation's 100 Resilient Cities, and Miami-Dade County's Resilient305. A discussion follows of other programs and initiatives on the county level, for Miami-Dade, Broward, and Palm Beach counties, as well as what the main cities in the region are doing to address climate resilience and community nutrition resilience. Attempting to provide a complete picture of actions in the region, the chapter then examines what key institutions and organizations are doing to address these challenges.

Keywords Miami · Southeast Florida Regional Climate Change Compact · Rockefeller Foundation's 100 Resilient Cities · Resilient305 · Resilience actions · Community nutrition resilience actions

The rapid population growth driven by migration and immigration, and consequent developmental expansion of the region, have led to several systemic stresses in Greater Miami. Among those are a shortage of

© The Author(s) 2020
F. Alesso-Bendisch, *Community Nutrition Resilience in Greater Miami*, Palgrave Studies in Climate Resilient Societies,
https://doi.org/10.1007/978-3-030-27451-1_3

affordable housing, transportation challenges, and a lack of well-paid jobs, as previously discussed.

The warming climate exacerbates the challenges to livability in Greater Miami, particularly through sea level rise (SLR), risk of flooding, beach erosion, storm frequency and/or intensity, urban heat waves, and consequent human health impacts. Federal, state, and local governments have begun taking action on these changes.

This chapter is looking at the actions of Greater Miami on resilience overall and community nutrition resilience in particular. It thereby starts with a brief discussion of actions driven by the federal government and State of Florida, then talks about actions on county level, and finally introduces city actions. Best efforts have thereby been made to capture the most important initiatives; however, due to the rapid emergence of new initiatives, not all might have been captured.

The definition of resilience in Greater Miami has been found to follow that proposed by Rockefeller's 100RC as "the capacity of individuals, communities, institutions, businesses, and systems within a city to survive, adapt, and grow no matter what kinds of chronic stresses and acute shocks they experience" (100Resilientcities.org 2018). Although officials and some institutions such as the Beacon Council talk about economic resilience as one of the cornerstones of resilience in Greater Miami, resilience actions are primarily focused on environmental, with the social pillar entirely missing from discourse at the time of this publication. Due to its natural challenges of porous limestone underground and the consequent threat of SLR to valuable assets as discussed in Chapter 1, resilience efforts in Greater Miami are mainly focused on adaptation to rising sea levels.

To better analyze the actions being taken, 25 interviews were conducted with representatives of local government, key institutions, and organizations working with climate resilience and community nutrition resilience. A full list of interview partners can be found in Appendix A. Findings have been added to the proceeding discussion.

Omitted in this discussion have been any actions dealing with sustainability, including the mitigation of greenhouse gases (GHG) emissions and otherwise preservation of natural resources, as this book aims to focus on resilience, i.e., the response to chronic stresses and acute shocks. Equally, the discussion will focus on long-term efforts; hence emergency services will not be looked at.

3.1 Political Action

Given the environmental, economic, and social impact of climate change on the world, the USA and South Florida, there is a clear discrepancy between what should be done to ensure community resilience and how fast, and what is being done. One of the causes of this lack of consolidated, impactful action is the lack of leadership on political level. Without this leadership, local policy and change makers oftentimes lack critical support and financial resources to improve the food security and nutrition security status of communities. Below follows a brief discussion of political action on federal, state, and local level.

3.1.1 Federal

During the interviews with elected officials and resilience professionals in Greater Miami, it became apparent that resilience efforts in the region are hindered, or at least made more difficult, by the unfavorable political environment on US federal level.

Historically, there has been a lot of miscommunication around the existence and causes of the warming climate, arguably supported by the fossil fuel industry. This has made it increasingly complex for people to understand the issue. In 2018, only 60% of people in the USA believed in climate change and that it is caused by human activity (The Hill 2018) (compared with about 90% in Western Europe). This has led to stalling of action on governmental level and arguably to the USA, losing a lot of time to take action.

The current administration under President Trump most certainly lacks leadership in this area and even withdrew from the Paris Climate Agreement, the global landmark agreement to combat climate change and to accelerate and intensify the actions and investments needed for a sustainable low carbon future (UNFCC.int 2019). However, regardless it is expected that the USA is going to achieve the emission reduction goals because of the work of Cities, driven by their mayors and by the private sector. Moreover, despite or perhaps because of President Trump's lack of commitment, initiatives have been proposed by other parties.

Overall, interview partners feel that the federal Government's functionality is reduced to little groups that are still functioning since the previous administration as for instance the National renewable energy lab.

These "pockets of action" are still operating, although off the radar. Although currently real progress is stalled, with no new people or committees being appointed, there are still people in the government that are continuing work on climate action.

An example of these pockets of action includes, for instance, the Climate Solutions Caucus. The caucus is a bipartisan group in the US House of Representatives which will explore policy options that address the impacts, causes, and challenges of our changing climate. The caucus was founded in February of 2016 by two South Florida representatives Rep. Carlos Curbelo (R-FL) and Rep. Ted Deutch (D-FL), who serve as co-chairs of the caucus. Membership will be kept even between Democrats and Republicans (Citizens' Climate Lobby 2018), which to many is a litmus test that the gap between parties related to the topic climate change is narrowing.

More recently, the Green New Deal was proposed by members of the Democratic Party, which is a set of economic stimulus programs that aim to address climate change and economic inequality. The name refers to the New Deal, a set of social and economic reforms and public works projects undertaken by President Franklin D. Roosevelt in response to the Great Depression. The Green New Deal combines Roosevelt's economic approach with modern ideas such as renewable energy and resource efficiency.

And there are other individuals on federal level that support sustainability and/or resilience-related efforts. For example, Carlos Curbelo, a Republican who served as the US representative for Florida's 26th congressional district from 2015 to 2019, has been mentioned by many of the interviewees as an example of someone in Congress that supports climate action. He, for example, proposed a carbon tax in 2018 as a measure to cut carbon emissions. He is currently considering a run for Mayor of Miami-Dade County (MDC) (Miami Herald 2018a), which could introduce new commitment to climate change mitigation for the county.

Related to resilience and particularly climate adaptation, MDC was recently selected to be the local sponsor for a US Army Corps of Engineers study on coastal storm flood risk management in the back bay or inshore waters surrounding Miami-Dade. This three-year US$3 million study is an opportunity for our county to make recommendations to the Corps of potential projects that would protect the county's infrastructure and people. These projects may receive federal funding in the future (US Army Corps of Engineers 2019).

When looking at federal support to enhance community nutrition resilience, established programs currently do not particularly consider disruptions due to severe weather caused by a warming climate. The two primary federal nutrition assistance programs in the USA that subsidize food purchases are the Supplemental Nutrition Assistance Program (SNAP) and the Special Supplemental Nutrition Assistance Program for Women, Infants and Children (WIC). Both programs are managed by the US Department of Agriculture (USDA).

3.1.2   State

On state level, progress has been slow during the past seven years under Gov. Scott's administration. State employees in the past have said that former Gov. Scott's administration warned them not to use the phrase climate change in state documents (Miami Herald 2018c).

Traditionally a purple state, Floridians cherish the environment for their personal enjoyment and for the benefits to the local economy. Especially, the critical tourism and real estate industries benefit from the beauty of local beaches and landscapes; hence, the need to protect it is high. Moreover, almost every Floridian has firsthand experience with flooding or tropical storms. Hence, there is awareness of changes in the climate although often it is not understood by what these changes are caused.

Resilience professionals in Greater Miami try to stay optimistic about where support for climate change action from the state is going. According to interview partners, state officials are waking up to the fact that more support is critical. Overall, interviewed professionals feel that better policy at federal and state level is needed, as well as financial support and technical assistance for critical infrastructure improvements. However, even with support from state-level missing, partnerships such as joint satellite-imagery and data collection are happening. It remains a challenge for professionals working in the space to better scale the few projects and financial resources from federal and state level.

3.1.3   Greater Miami

Even with the lack of support from federal and state level, local governments have significant responsibilities to its residents, as well as the capabilities to take action. Dan Gelber, Mayor of Miami Beach, in the

2018 South East Florida Climate Action Leadership Summit summarized the sentiment among local government by saying "We cannot draw a line. Water doesn't draw a line, and neither does resilience."

However, more can be done. Some resilience professionals feel that collaboration between the 35 local governments has been slow. Further, it is perceived that action on county level, particularly in MDC lacks a clear direction, so that resources and needed support can be provided. Jane Gilbert (2018), Chief Resilience Officer (CRO) of the City of Miami hopes that with updated Resilient305 strategy, published on May 30, 2019, more and quicker action can be taken. Even if this doesn't include financial support, she expects better results from shared data and common guidelines.

Philip Stoddard, Mayor of South Miami, summarizes the sentiment of many of the interviewed individuals working on both sustainability and resilience in Greater Miami with the realization "I'm it. There are no partners upstream." Several times during the interviews for this book it was mentioned that given the lack of support, professionals operate within their pockets of power, focusing on what they can do on a local level.

3.2 REGIONAL RESILIENCE ACTION

Greater Miami's resilience efforts have been happening mainly under three umbrellas: the Southeast Florida Regional Climate Change Compact, Rockefeller Foundation's 100RC, and Resilient305. These will be briefly discussed below.

3.2.1 *Southeast Florida Regional Climate Change Compact*

After extensive stakeholder consultation, the Southeast Florida Regional Climate Change Compact was executed in January 2010 to coordinate climate change mitigation and adaptation efforts across county lines in the four counties: Broward, Miami-Dade, Monroe, and Palm Beach. It has been hailed as one of the premiere multi-jurisdictional climate efforts in the USA and was cited by President Barack Obama "as a model not just for the nation, but the world" (Southeastfloridaclimatecompact.org 2017).

In collaboration with municipal partners, key strategies are attempting to create more economically competitive and climate-resilient communities are being implemented. The Regional Climate Action Plan (RCAP) includes 110 action items aiming to reduce GHG emissions and adapt to the effects of climate change until 2020 (Southeastfloridaclimatecompact. org 2017).

According to Carlos Gimenez (2018), Mayor of Miami-Dade County, the achievements of the Compact until 2018 have been manifold. First, it successfully convened scientists to develop unified SLR projections for South Florida (Resilient305.org 2018) until 2100. Secondly, it provides a tool to focus on reducing GHG emissions. For Miami-Dade, it has led to an agreement with FPL to install over 1 million solar panels over the next three years. Next, it critically enabled partnerships across the four counties as well as cost-sharing for critical studies and actions. It also added a focus on economic resilience with its Economic Resilience Committee in collaboration with the chambers of commerce. MDC's economic development and resilience work are done through the Beacon Council and its "One community-one goal" program, which is the name for the long-term strategic plan for Miami-Dade County's future economic development success (Beacon Council 2019). Finally, the collaboration in the compact enables the counties to speak with a united voice on state level to request support and funding.

However, the RCAP does not cover food system or community nutrition resilience. It has two recommendation areas under which it would fit. The first is Public Health, the goal of which is to build capacity to proactively mitigate climate-related public health risks in Southeast Florida (Southeastfloridaclimatecompact.org 2019a). However, it currently only covers risks related to heat-related illnesses, and chronic conditions such as asthma or diabetes, as well as to pathogens and mosquito breeding due to a proliferation of flooding. The second area of recommendations under which community nutrition resilience would fit is Agriculture. However, looking at the recommendations they appear to focus more on the goal of ensuring the continued viability of agriculture in Southeast Florida in the face of climate change (Southeastfloridaclimatecompact.org 2019b). They include recommendations on policies and actions that encourage sustainable production, remove barriers to production, promote economic incentives, improve water reliability, and promote best management practices. One recommendation that is related to complimentary food access points is the goal of urban agriculture. It covers identifying and reducing obstacles to enable urban agriculture, gardening, and other backyard agricultural practices.

3.2.2 100 Resilient Cities (100RC)

100 Resilient Cities (100RC) is a US$100 million effort launched by the Rockefeller Foundation in 2013 to help cities build resilience to social, economic, and physical challenges with the aim of enabling them to better respond to impacts of urbanization, globalization, and climate change. The 100RC network comprises about 112 cities spanning 40 countries and 27 languages, one of them being Greater Miami and the Beaches (GM&B) (Resilient305.org 2018). With 2.7 million residents, Miami-Dade County, the City of Miami Beach and the City of Miami make up what is commonly referred to as GM&B. GM&B is composed of 35 local governments.

In 2016, GM&B was selected to join 100RC as the first collaboration between a county and two cities. As part of this membership, GM&B has been since provided with resources to develop a roadmap to resilience through:

- Financial and logistical guidance for establishing the position of a CRO in the City of Miami;
- Expert support for development of a robust resilience strategy;
- Access to solutions, service providers, and partners from the private, public, and NGO sectors who can help them develop and implement their Resilience Strategies;
- Membership of a global network of member cities who can learn from and help each other.

According to Torriente (2018), before joining 100RC GM&B's resilience work was focused on SLR. With the membership, she argues, this focus has broadened, now including economic and social resilience. Food security is featured as one of the challenges that the resilience strategies of 100RC member cities are addressing. However, GM&B has not identified food security as one of the challenges they will be working on since it wasn't prioritized by consulted stakeholders. However, the ways in which other cities address this challenge can inform how GM&B could address it in future. Hence, a discussion of the actions of selected cities to improve food security will be discussed in Chapter 4 of this book.

3.2.3   Resilient305

In 2016, after joining 100RC, GM&B formulated their Resilient305 initiative to respond to GM&B's main resilience challenges. The strategy was formulated after an extensive phase of listening to and learning from stakeholders which included an online questionnaire with over 2000 responses, 14 interactive focus groups, workshops and meetings with mayors, and nine subject-specific meetings. Based on this in 2017, planning for the resilience strategy began.

Stakeholders were consulted around six key issues: housing, transit, water resources, education and youth development, and health services. The priority shocks and stresses that were identified in meetings with the general public, focus groups and Mayors included the transportation system, which is perceived as overtaxed and unreliable, a lack of affordable housing, aging infrastructure and rising sea levels and coastal erosion (Resilient305.com 2018). Stakeholders stated in the questionnaire that climate actions, urban mobility, and the environment and sustainability should be prioritized in GM&B's 20-year resilience strategy (Resilient305.com 2018).

On May 30, 2019, an update to the Resilient305 strategy was presented. Under the three pillars Places, People, and Pathways, it now includes 59 actions to address GM&B's key resilience challenges (Resilient305 2019). Objectives under the "People" pillar include to cultivate financial stability, advance public health priorities, strengthen community response, and communicate the concept of resilience. Notably, most of the "People" actions are largely unfunded or funded by partners such as the Health Foundation of South Florida, United Way or the Miami Foundation instead of county or city budgets.

Food and nutrition are not part of either the past or the current Resilient305 strategy. And that is despite stakeholders during the consultation for the first strategy (under the key issue "Health and Wellbeing"), identified access to healthy, affordable food as an area where GM&B needs to improve. Further, healthy habits and access to healthy food choices came third in terms of top health and safety priorities in the questionnaires. Moreover, in the event of a disaster like a storm, "getting power, food and water" was ranked top priority.

The following summarizes the actions taken by the major cities in the three counties Miami-Dade, Broward, and Palm Beach, with resident numbers above 100,000. Summaries are based on conducted interviews, and on secondary research including the submission of cities to the RCAP, summarized in Appendix B.

3.3 MIAMI-DADE COUNTY

Reportedly, MDC is spending US$600 million in 2019 in operating expenses on climate change mitigation and resilience, including monies for SLR related infrastructure projects. It is also flagging US$16 billion in unfunded projects for SLR related infrastructure upgrades until 2024 (Florida Phoenix 2019).

In 2015, MDC hired Chief Resilience Officer Jim Murley, who now leads a team of 11 people in the Office of Resilience and Sustainability. The most active cities in MDC are the Cities of Miami, Miami Beach, and South Miami. Notably, Miami Gardens is running a Live Healthy Miami Gardens initiative. However, Hialeah with 225,000 residents (the sixth-largest city in Florida) appears presently not to be doing anything in the field.

On SLR and climate change action, the county is still very much in planning and preparation mode. They are currently performing a vulnerability assessment of key infrastructure and capital investments. The MDC Water and Sewer Department is investing billions on improving critical infrastructure, and SLR is a key part of the design criteria. They are also investing in protecting the county's drinking water and actively monitoring and managing any saltwater intrusion into the aquifers. Further, a Sea Level Rise Task Force was created. Moreover, several investigative activities such as in collaboration with the South Florida Water Management District (SFWMD), the Rand Corporation, or the Urban Land Institute (ULI) are ongoing. In order to protect and restore the Everglades, MDC works with the SFWMD and the US Army Corps of Engineers on implementing the Comprehensive Everglades Restoration Plan. The county also adopted a Green Sustainable Building Ordinance for new construction (Resilient305.com 2018).

In July 2006, the MDC Board of Commissioners established the Miami-Dade Climate Change Advisory Task Force comprised of 25 appointed members representing various sectors of the community. The task force is charged with identifying potential future climate change impacts to Miami-Dade County and providing ongoing recommendations to the Board of Commissioners regarding mitigation and adaptation measures to respond to climate change (Miamibeachfl.gov 2018). The county summarizes its activities in an interactive map for residents to inform themselves (Mdc.maps.arcgis.com 2018). It also runs courses to teach residents about SLR.

In addition, MDC is developing its "GreenPrint: Our Design for a Sustainable Future." This document will be the framework to evaluate and integrate environmental, social, and economic benefits into county policies and initiatives (Miamibeachfl.gov 2018).

With regard to community nutrition resilience, there are 640,000 residents that are SNAP eligible in the county, hence food insecure. Some of the initiatives that the University of Florida (UF) IFAS Extension program is working on to improve nutrition resilience in GM&B include the Family Nutrition Program, the Supplemental Nutrition Assistance Program Education (SNAP-Ed) program in the State of Florida, which encourages people to make a better choice. Further, they work on converting food pantries to choice pantries, where people are given the dignity of choosing foods themselves instead of getting a pre-selected box. This increases the cultural relevance, and thus consumption of the healthy food items in the pantry.

Looking at resilience in the food system, MDC created an Agricultural Manager position in 2005 to facilitate collaboration between the county government and the agricultural industry. Some of the initiatives run by this position are a tax-payer supported bond program to buy development rights on farmland, code enforcement, as well as policy issues (federal and local) that affect the industry, including branding, labeling, and land use.

The main programs that MDC is doing to promote food security include labeling. The "Fresh from Florida" brand aims to ensure that people understand the significance of farming on local level. Most farmers and packers in MDC have signed up to use these labels. "Redland raised" is another more local brand that is part of Fresh from Florida. MDC has supported the brand with marketing in stores until there was no more budget. Since then, growers in MDC have started to put on all products, in-store, on boxes and it has helped promote local agriculture.

In addition, the county tries to create alternative market places to keep farmers in business. These include the promotion of local food and flowers for destination weddings or farm weddings, food tours or pick your own's. The county facilitated these by limiting the liability of the farmer which helps with signage and insurance premiums.

There is a variety of programs helping to support and/or promote local agriculture. For example, the Homestead-based Redland Farm Life school project, originally built in 1916, and recently refurbished, now teaches mechanic work, cooking and home economics at a middle and elementary school. Further, according to Lapradd (2018), a culinary school is planned where food training shall be provided. It will also include a commercialization center and a state-certified kitchen where local food entrepreneurs can create new products.

On consumer level, the Jesse Trice Community Health Center, which is one of Miami-Dade County's preeminent federally qualified community healthcare centers offers a **Fruit and Vegetables prescription** program. Supported by the nonprofit Wholesome Wave, the center's doctors have been giving prescriptions to families that are food insecure and suffer from dietary-related diseases. These can then pick up the produce at the health clinic when they get their checkup done. They can also attend educational events such cooking classes at the center.

Another noteworthy program is the Consortium for a Healthier Miami-Dade, which is sponsored by the Florida Department of Health and works toward healthy environment, healthy lifestyles, and a healthy community (Healthymiamidade.org 2019).

3.3.1 City of Miami

The City of Miami is a member of the Milan Food Policy Pact, which means that they are committed to building sustainable, equitable, and resilient food systems. Jane Gilbert heads up the Office of Resilience and Sustainability which consists of 3 full-time staff. Originally funded by the Rockefeller Foundation, it is expected that Jane's position will be extended once the funding runs out.

In an interview with the radio station WYNC, Mayor Francis Suarez recognizes that the term resilience is broader in scope than what the City of Miami is currently focusing on which is mainly the prevention of flooding and the adaptation to SLRs (Wnycstudios.org 2018). Economic resilience needs to be in scope as well, as well as protecting communities from climatic gentrification. This is echoed by Mayor Philip Stoddard of South Miami as economic resilience can only be achieved through a stable tax base of people, who are able to live in a livable city.

Funding for resilience measures was secured in 2017, when City of Miami voters passed Miami Forever, a US$400 million general obligation bond. About half of that will help Miami adapt to SLR, including projects such as storm drain upgrades, flood pumps, and sea walls. Another US$100 million will help more Miamians get access to affordable housing although it is not yet clear whether this money is going to address resilience-related issues such as the installation of AC units.

The remaining US$100 million will go to projects like improving public spaces and roads, and public safety (Miami Herald 2017). Food system resilience is not being addressed.

In order to improve community resilience, the City of Miami is working on different fronts. For example, single-family houses and rental properties in affordable housing are being retrofitted for energy efficiency and to make them more prepared for severe weather, such as with hurricane shutters. The city is also working on the issue of transport and mobility with a bicycle master plan, new bus lanes and shades for stops of trolleybuses to foster uptake of public transport (Gilbert 2018).

Actions on SLR are guided by a SLR Advisory Committee, established in 2015, and an interdepartmental Resilient Infrastructure Committee, founded in 2017. These have guided several actions including the update of the city's stormwater master plan as well as stormwater upgrades in highly vulnerable areas. Further a rapid action plan for flood risk mitigation of critical infrastructure. Flood risk mitigation was also strengthened in the Future Land Use and Coastal Management elements of the city's Comprehensive Neighborhood plan. Moreover, the city's zoning code contains several standards aimed at maximizing natural infiltration of stormwater directly to the ground (Resilient305.com 2018).

Education and communication with the public remain a critical part of the action plans. Though SLR is a constant topic of discussion and debate in South Florida, the average resident in Miami might not really understand how it impacts them. The city hopes to change that with a US$100,000 grant from the Bloomberg Philanthropies' Mayors Challenge. It'll use the money to develop an interactive tool that'll highlight sea rise predictions, the impacts of infrastructure improvements and other weather-related data sets (Miami Herald 2018b).

Food system resilience has not been addressed directly. Examples of what the City is doing include a Resilience and Equity checklist for residential developers, which includes among other checks that there is a supermarket access nearby. Further, transit-oriented development guidelines aim to address the nexus between housing, healthy food access, and transport, e.g., for mixed-income/mixed-use communities. They also recommend having transit stops close to stores as a zoning requirement. Thirdly, Community Operations Centers that should act as "Resilience hubs" and could in future include a food element such as cooling hubs from where food can be distributed post disruption. However, more robust work needs to be done to understand food availability systems.

3.3.2 City of Miami Beach

The Miami Beach Office of Resilience and Sustainability is headed up by Susanne Torriente. Recently, the City hosted the United States Conference of Mayors and is internationally recognized for its adaptation projects (Resilient305.com 2018).

Resilience efforts go back to 2009 when the city established a Sustainability Committee and codifying Chapter 100: Sustainability in the City Code, which is dedicated to sustainable initiatives (Miamibeachfl.gov 2018). Since 2013, the city has taken steps to protect its investments from the impending effects of climate change. Miami Beach is now approximately 15% into a ten-year, US$600 million multiyear stormwater management program that addresses both land use and development code and infrastructure updates. This resilience plan, dubbed "Miami Beach Rising Above" by city officials, includes improving drainage systems; elevating roads; installing pumps to replace aging stormwater pipes; replacing much of the city's water, wastewater, and utilities systems; and updating regulations to reflect increased elevation requirements, seawall barriers, and more (Urban Land 2018).

To hold back the water, the City of Miami Beach has made various infrastructure improvements. Pumps have been installed that force water back into Biscayne Bay in parts of the city known to flood. Further, some of the seawalls have been reinforced and valves put in pipes to prevent water from flowing back up. There are about 30 pumps installed so far, with up to 90 planned in total (Business Insider 2018). The city also raised roads in compromised, highly trafficked neighborhoods like Sunset Harbour, West Avenue, and the Flamingo Historic District (Urbanland.uli.org 2018). Further, the City also enhanced its disposal process to ensure water quality, adding water-treatment stations so that the water being pumped into the bay is cleaned from debris that pools in the streets when storm drains get backed up (Business Insider 2018). It also incorporated green infrastructure to further mitigate against the effects of SLR while also enhancing the community. The city financed the majority of these efforts by selling off US$100 million worth of bonds earmarked for capital improvements, coupled with public funds derived from Miami Beach's substantial tax base (Urbanland.uli.org 2018). In addition, Miami Beach voters approved a US$439 million general obligation bond in November 2018.

Revamping building codes to allow developers to build higher is another way to deal with the effects of SLR. In a place like Miami Beach, there's resistance to elevating structures and blocking views since everyone wants to see the beach, but that resistance is incompatible with making the adaptations needed (Business Insider 2018). In April 2018, a ULI Advisory Services Panel concluded that the city has made an admirable start on resilience efforts, but a more comprehensive and holistic approach needs to be taken (Urban Land 2018).

Climate literacy is another challenge the city is addressing. The city's communication plan to roll out this effort included a branded Rising Above Web site, a social media campaign, and extensive community outreach to local organizations. When talking about resilience, many people in Miami Beach still think about stormwater. One of the strategies for the updated Resilient305 strategy is to have improved education, communication strategies and tools such as graphics, art pieces, and cultural events (Resilient305 2019). It is also recognized that different resilience communication is required in different neighborhoods.

Food and nutrition security is currently not addressed. However, Torriente agrees that these areas must be developed more especially since Miami Beach as an island community is vulnerable to disruptions in the case of severe weather. Regarding systemic food system resilience, Miami Beach is a community of great wealth, although it has some pockets of poverty in the North, hence the issue of food deserts is not as pressing as in other cities of Greater Miami.

3.3.3 Miami Gardens

Miami Gardens, the third-largest city in MDC, runs a noteworthy program to improve the health and well-being of its communities, the Live Healthy Miami Gardens program. With a predominantly African American population, 76%—the 3rd largest in the USA by population, it includes 34,000 households. First efforts were made in 2014, when a General Obligation Bond of US$60 million was issued to rebuild and/or renovate the City's parks.

A 2014 survey with the community revealed that health disparity, and the lack of physical activity was the number two priority issue for the community after safety, and number one for children. In the elderly, obesity was cited as a particular concern. There are two food deserts in the city. Also in 2014, the Health Foundation of South Florida selected the City of Miami Gardens for a 6-year community health grant. Healthy eating and nutrition were selected as two of the focus areas. The project also got awarded a National League of Cities, CHAMPS grant of US$125,000, as well as the CDC REACH grant of US$600,000 per year for 5 years. Some of the strategies to increase access to healthy food include mobile farmers markets, a healthy restaurants initiative, a healthy convenience store initiative, community gardens, as well as permanent farmers markets in the future.

3.3.4 South Miami

The City of South Miami stands out for its progressive action on sustainability and climate change mitigation. Mayor Philip Stoddard is personally interested in climate resilience due to his background as a biologist and expert in disturbance ecology. During his tenure, he has implemented several noteworthy programs to address climate change mitigation and resilience. Among these are an ordinance that requires solar panels on every residential new build. This is beneficial for both economic and environmental reasons, as realtors can sell the house faster and homeowners make a positive return due to their energy cost savings (Stoddard 2018). The ordinance generated interest from the media internationally.

As an elected official, Mayor Stoddard has been jumpstarting new ways of doing things including bringing farmers markets and urban gardens (garden in the front yard) to the City. He does so by teaching model behavior in his city and private life (his own house is run by solar power). However, his priority with the resilience program is to stabilize the City's tax base by ensuring future livability in South Miami.

Community nutrition resilience and food system resilience currently not featured in planning in South Miami. The only programs that are running such as farmers markets and 2 for 1 SNAP benefits for fruits and vegetables are led by grassroots organizations.

3.4 BROWARD COUNTY

According to Broward's Chief Resilience Officer Jennifer Jurado (2018), Broward County's resilience work today is benefiting from an early investment into climate change adaptation, particularly related to SLR and water management. Jurado has been involved in water management and ecosystem restoration on county level since 2000, when she was asked by the county to create their first task force on climate change. Hence, the support from leadership for both sustainability and resilience in Broward County is strong.

The county's priorities include the update of the advanced future condition maps series. This series represents the expected *future* average wet season groundwater elevations for *Broward County*. The first map in the series thereby influences drainage and stormwater requirements and is expected to initiate change to the current land use plans with the aim of establishing regionally consistent seawalls. Moreover, the county is currently updating the 100-year flood maps that will include SLR and rain projections, as well as groundwater increase. The goal of this initiative is a full evaluation of vulnerabilities related to the current infrastructure and functionality of systems, both temporarily and spatial. Although the methodology of the flood maps update is being developed in Broward, with the help of academic partners, it can and will be employed outside of region and will thus impact flood insurance premiums in the region.

A further goal is to integrate the South Florida Flood Control project with the Coastal study done by the Army Corps of Engineers, as well as find other opportunities to benefit from federal funding. Based on these findings, there's also a need to organize resources to conduct capital improvements.

Broward County is collaborating with its municipal partners in several ways. For example, 10 cities provide cost-share funding in 100-year flood map program. Those that do will gain discounts for residents for flood insurance programs, as temporarily they can preserve their rating in FEMA.

The county further has been helping cities in conducting vulnerability assessments for coastal communities, has held Resiliency round tables and is regularly conducting workshops with the cities' resilience points of contacts. The county also works on achieving consistency in data and actions across cities. It helps with evaluating local data and provides technical assistance that is helpful to cities. Since there is not enough funding on neither county level nor city level, efforts are being shared and often the county develops something and offer cities the option to opt-in, which most do.

With regard to community nutrition resilience, Broward County has been working on efforts to improve community food security and nutrition resilience for the past 5 years. In 2014, Broward County hosted the first workshop on Food Systems Planning and Food Rescue. The event was organized for the Sustainability Stewards of Broward (SSB) group, a peer-exchange with over 450 members who share best practices and collaborate on policy and program development in person and online. More than 45 local planners, farmers, health professionals, business leaders, and elected officials joined the workshop, which was a first conversation about food deserts and food insecurity from an environmental planning perspective.

In 2016, Broward County shared their Green Infrastructure Map Series for the SSB network focused on Local and Regional Coordination on Green Infrastructure. It was the first time the maps were released to the public and participants were invited to give feedback on the consideration of Habitat Corridors, Energy, and Food Systems in the maps.

In 2017, Broward's Environment Planning and Community Resilience Division launched the Corporate Sustainability Network (CSN) for leaders in the business community focused on sustainability. Over 60 individuals, representing local businesses, national corporations, and state and local government agencies, attended the kick-off event to discuss present challenges around 3 main areas of sustainability: Renewable Energy & Energy Efficiency, Waste & Recycling, and Green Hospitality & Food Recovery.

Already in 2014, the county had included a climate change element in their comprehensive plan, which lays out how building can happen in the county and under which rules, including conservation and housing. This element was updated in 2017 with a strong focus on food policy and actions on, e.g., food deserts, community gardens, and composting. Though the comprehensive plan is reviewed by the state and constitutes a strong recommendation from the county to municipalities, they don't have to follow it. However, in 2017 the focus on food policy was also passed by commissioners to be included in the land use plan, which is mandatory for municipalities to follow. The final draft of the Broward County Land Use Plan now reflects the holistic view of green infrastructure and includes recommendations from the initial workshop on Food Access, resulting in the first policy citing Food Deserts as a community concern (Broward.org 2019).

Another noteworthy program in Broward County is People's Access to Community Horticulture (PATCH™), a community gardens program. Through the creation of a Community Garden on a former trash dump, it aims to improve the quality of life in the urban residential areas in Dania Beach promoting therapeutic activities, strengthen connections in the neighborhood and providing access to locally grown food (Thepatchgarden.com 2019).

Broward County seems to be leading the way on food recovery in the region. As one of the outcomes of the CSN, a Food Recovery Workgroup was created which continues to meet bimonthly and make progress in expanding Food Rescue and Recovery efforts through Broward County schools and corporate partners in the hospitality industry through education and logistical support.

Food waste avoidance and food recovery serve as a climate mitigation and adaptation strategy. According to the Map the Meal Gap 2018 report by Feeding America, 11.8% of South Florida's population is currently at risk of hunger. Meanwhile, over one billion food items are wasted annually in schools in the USA (Food Rescue 2019). While Broward County public school cafeterias are very efficient, the community realized there might be opportunities to rescue extra food to benefit local food banks, while creating a teaching moment for students about hunger and mindful consumption.

In 2017, Broward County staff and CSN partners began a conversation with the Food Services Director for Broward County Public Schools. Despite federal and state good Samaritan laws like the Bill Emerson Act, principals were hesitant to create a food recovery program without a policy from the school board. A few months later, advocacy efforts of students and teachers resulted in two schools, Beachside Montessori Village and Marjory Stoneman Douglas High School, launching food recovery pilot programs for the district. Teachers at Beachside Montessori Village, a public magnet school, partnered with foodrescue.net, Kids Can, and Broward Outreach Center to pilot the first major food recovery program in a Broward County school. As part of this program, students are encouraged to donate their leftover unopened, uneaten, or unpeeled food items. The school's sustainability class, which studies the relationship between food waste and its impact on climate change, keeps track of the amount of food items rescued to measure the impact they are making on the environment. So far, over 1000 items are donated to a local food bank every week, keeping the equivalent of 132 tons of carbon dioxide out of the atmosphere. Marjory Stoneman Douglas High School is also working with Broward County Public Schools to collect unwanted, unopened, and nonperishable items from students' lunches. Spearheaded by a sophomore who wanted to make a difference in her community, the project is now a role model for the entire district. The program is simple and effective. Every day, an announcement is made at the end of lunch, and items are collected in a donation box. The project benefits Gateway Community Outreach, which serves the homeless (Horwitz 2019).

The programs at Beachside Montessori Village and Marjory Stoneman Douglas High School collaborated with each other and the district to discover best practices as they went through their pilot year. So far, the initiatives are considered successful, and district staff are incorporating local health department rules into new policy guidelines. With guidelines in place, full-scale replication throughout Broward County Public Schools was rolled out Fall 2018. So far, 15 schools have signed up to launch a program in the 2018–2019 school year; and on August 10, 2018, Beachside Montessori Village, which currently leads the effort, hosted a training for schools interested in implementing a food recovery program. Each school will need an extra refrigerator to sort and store donated food, so funds are being collected through CSN partners and the community-at-large. These efforts support progress toward the goals in the county's Climate Change Action Plan in several sections, especially those related to resource efficiency, waste reduction, greenhouse gas reduction, local food systems, and community outreach. With over 200 schools in the district, Broward County has the potential for offsetting over 1.5 million pounds of GHG each year.

At the same time, CSN partners are supporting expansion of Food Recovery/Rescue throughout the hospitality industry in Broward. Green Meeting Industry Council's Food Rescue Committee is leading the charge, with support from Broward County's Environmental Planning and Community Resilience Division, the Florida Department of Agriculture and Consumer Services, South Florida Hunger Coalition, Florida International University, and others, to spread the message and train new chefs and event managers how to plan, contract, and run a successful food rescue. For example, food recovery at the Southeast Florida Regional Climate Change Compact Summit in 2018 resulted in over 900 pounds of prepared food rescued and delivered to Life Net for Families and over 600 attendees were exposed to the project. On May 16, the Sustainable Events Network they worked with Broward County Convention Center to rescue 39 trays of food from a huge event called Biz Bash, which was sent to the Broward Outreach Center. At the same time, the chef realized an opportunity to use the donation and truck to clean out the fridge and cupboards of over 1000 pounds of food that was soon to expire. Currently, the partners are working to negotiate a contract with Food Connect to launch a free community-wide App connecting donors, drivers, and receivers. All of these efforts are supported voluntarily, with no funding or paid staff time available from the parties.

3.4.1 Fort Lauderdale

Fort Lauderdale has been working primarily toward creating a more sustainable city for many years. Its Sustainability Division was formed in 2012 to consolidate resources toward this goal. Under the umbrella of sustainability, this division manages public works not only to provide basic service or comply with the law, but also to reduce negative environmental impacts, to maintain the livability within the city, and to reduce the costs of these operations through innovative best green management practices. One of the primary responsibilities of the division is implementing, monitoring and updating of the city's Sustainability Action Plan (SAP). The SAP provides a means for articulating: (1) the city's specific "green" goals, strategies and performance indicators, (2) how sustainability will be integrated into all levels of City decision-making, and (3) a system of accountability. The initial SAP was published in 2010 through a collaborative effort of the Sustainability Advisory Board, City staff, expert groups and concerned neighbors. The final result considered what was working, what was missing and how each effort helped or hindered the others, was a coordinated statement introducing five priorities and 17 goals. There were eight goals including Leadership, Air Quality, Energy, Water, Built and Natural Environment, Transportation, Waste, and Progress Tracking. The SAP identifies new initiatives to encourage and assist residents, businesses, developers, and others to practice sustainability.

Long-term and short-term risks are being considered by the City of Ft. Lauderdale in future planning scenarios. The city is using a geographic information system (GIS) to predict and map potential environmental disasters, as well as during emergency response (ICMA.org 2018). The long-term risks include outdated flood control structures, saltwater intrusion and impacts of tidal flooding. The city is addressing these threats through the update of flood control structures to current

SLR projections (Ces.fau.edu 2014). Short-term, the risk of storm surge is being addressed through adaptation of infrastructure and surge protection structures, as well as in emergency planning.

A large-scale community outreach effort also formed the foundation of the 2035 Vision Plan, a plan meant to guide decision-making processes in the City (Fortlauderdale.gov 2018). The city tracks and publishes progress on the different areas of the SAP and 2035 Vision Plan. Further, Fort Lauderdale has been a partner in the Seven50 ("seven counties, 50 years") Consortium. This initiative has been developing a blueprint for growing a more prosperous, more sustainable Southeast Florida during the next 50 years and beyond. It is being developed to help ensure a vibrant and resilient economy, and stewardship of the fragile ecosystem in the mega-region (Ces.fau.edu 2014).

3.4.2 Other Major Cities in Broward County

Other cities in Broward County have been taking some resilience actions, too. The City of Hollywood is currently a municipal partner in the Southeast Florida Regional Compact. According to Jurado (2018), the city is very informed about the need for resilience action, and the mayor as well as the chamber is supporting climate adaptation efforts. Recently, the city issued an request for proposal for a small resilience study as well as a bond initiative to finance SLR improvements. They're also working on securing grant funds for resilience work.

The City of Miramar has adopted SLR projections of the region and is making an effort on resilience, according to Jurado (2018). In collaboration with the county, the city recently attempted an application for the Institute for Sustainable Communities to convene communities around sustainability and resilience.

Coral Springs is working closely with the county on updating the county flood maps though they are not yet a cost-sharing partner. Despite some residents wish they were doing more on sustainability, they have good investment on sustainability. Finally, due to its coastal location, Pompano Beach understands the necessity for adaptation work, particularly seawalls. The city has also been investing in one-way storm valves to deal with flooding.

Since Pembroke Pines has not been exposed to SLR as other communities have, there does not appear to be any noteworthy resilience work being done.

3.5 PALM BEACH COUNTY

Palm Beach County (PBC) consists of 39 municipalities that have been focusing on resilience to different degrees. The county hired Megan Houston as the Chief Resilience Officer in 2018 and she heads up a team of three people. According to Houston (2018), the Office is initially focused on developing a climate action plan including the tracking of GHG emissions, to promote PACE and a solar program, as well as adapting to SLR.

Related to community nutrition resilience, PBC has an extensive hunger relief program which was launched in 2015. Recognizing the unacceptable consequences of local hunger, United Way of Palm Beach County, the Palm Beach County Board of County Commissioners, backed by a group of 183 organizations, convened the Hunger Relief Project and identified the need to create a comprehensive plan to reduce local hunger. The Food Research and Action Center (FRAC) and the University of South Carolina (USC) Center for Research in Nutrition and Health Disparities were commissioned to create a Hunger Relief Plan in 2015 (Palm Beach County 2015).

The Hunger Relief Plan has ten goals, which are underpinned with objectives and strategies. These include goals to provide access to healthy, nutritious food to all PBC residents, partnerships between different sectors to make this happen and the increase of health literacy in the county. To implement this plan, all the stakeholders have been working together led by a Hunger Relief Executive.

The activities of PBC's main large cities West Palm Beach and Boca Raton are summarized below. In both cities programs, there appears to still be a higher focus on mitigation instead of resilience. Overall it can be observed that PBC overall is behind Broward County and MDC with regard to resilience.

3.5.1 West Palm Beach

The Office of Sustainability was created in the Public Utilities Department in 2008. In 2014, with Mayor Jeri Muoio's emphasis on the importance of Resiliency and Sustainability, the Sustainability Office moved into the Mayor's Office, providing increased access and involvement in City activities and initiatives (WPB.org 2018).

The main programs are 10,000 trees in 10 years, in which neighborhoods are provided with free trees by the city; e4 Life, an exposition at which residents can inform themselves about how to make environmentally conscious decision in their everyday life, such as to reduce their carbon footprint, or save water. West Palm Beach also was recognized with the national SolSmart Gold designation for its short, 1-day permitting of solar energy installation. Mayor Muoio, who has joined the "We are all in" network of leaders who support climate action to meet the Paris Agreement, also aims to transition the municipal fleet off fossil fuels by 2025.

3.5.2 Boca Raton

Boca Raton hired Lindsey Roland Nieratka as their first Sustainability Manager in May 2018 to coordinate the local resilience and sustainability initiatives. Current initiatives within the City can be found in each department and include LED fixtures in traffic signals, parking lots lights and street lights, electric vehicle charging stations, utility services' 100% reclaimed water system and sanitation's curbside recycling program. The City also has two boards dedicated to preserving and promoting a sustainable and environmentally friendly community: the Environmental Advisory Board and Green Living Advisory Board (Myboca.us 2018). Nieratka plans to increase community engagement around the topic. Moreover, Boca has started to discuss adaptation to SLR with the county (Houston 2018).

3.6 COMMUNITY ENGAGEMENT AND OTHER KEY INITIATIVES

Community engagement and activism is a key component of ensuring community nutrition resilience in Greater Miami. Given the lack of comprehensive planning and focused action of local government to respond to the complex challenge of community nutrition resilience, several communities have turned toward more localized initiatives, developing programs within which to pursue this goal. For example, after Hurricane Irma in 2017, elderly people in some neighborhoods went days without ice or water and some students who rely on free school lunches didn't have a way to eat. Volunteers and community groups stepped up to host barbecues, delivered supplies, and helped with debris removal.

This section discusses the key organizations and institutions that support community activism where systemic, government-led support is lacking to ensure nutrition resilience for Greater Miami's residents.

3.6.1 Key Nonprofit Organizations

There are many noteworthy organizations working on food security and nutrition in Great Miami. These include Feeding South Florida, a subsidiary of Feeding America, a nationwide network of more than 200 food banks that feed more than 46 million people in the USA through food pantries, soup kitchens, shelters, and other community-based agencies. Feeding South Florida plays a pivotal role in supplementing food supplies for food-insecure households. Locally, Feeding South Florida rescues and distributes 46 million pounds of food each year through direct-service programs and a network of nearly 400 nonprofit partner agencies throughout Palm Beach, Broward, Miami-Dade, and Monroe Counties (Feeding South Florida 2018).

Another organization with the same goal however an even larger focus on healthful food is FarmShare. This Homestead-based nonprofit, using inmate labor and volunteers, resorts, and packages surplus food collected from local farms and distributes it to individuals, soup kitchens, homeless shelters, churches, and other organizations feeding the hungry in Florida. It thereby redistributes 200% of the cost of goods sold back to the farmer, thus providing a win-win for all (Farmshare.org 2019).

Further, there is Urban Oasis Project, which connects people to local food through farmers' markets (stationary as well as mobile), farm dinners and urban gardens. In fact, Urban Oasis Project is involved in a variety of larger programs such as Live Healthy Miami Gardens, discussed earlier. The farmers market at Legion Park in Miami, which has been run by Urban Oasis Project for the past eight years, is probably the most well-known farmers market in the region, offers commercialization opportunities to local food businesses, and provides truly local food. It also accepts SNAP benefits and offers the double the benefits Fresh Access Bucks program (FAB) where a person on SNAP gets two for one on fruits and vegetables. Friedrich, President of Urban Oasis Project has been a champion of community nutrition resilience in MDC for the past nine years, having co-led the first MDC food policy council.

Urban Green Works (UGW), founded in 2010 by James Jiler, is a nonprofit focused on community food security and horticulture therapy (Urbangreenworks.org 2019). Their community-based environmental and food justice programs and projects include planting trees to increase canopy cover, to building food forest gardens to addressing food security issues in food desert neighborhoods. Originally founded to prevent crime through creating stronger attachment to neighborhoods as well as keeping kids off the street, UGW now is an important part of Greater Miami's urban ag movement.

Furthermore, there is Common Threads, a nonprofit which aims at preventing childhood obesity by teaching kids how to cook (Commonthreads.org 2019). They provide cooking classes to demonstrate people that healthy food can taste good.

There is also FLIPANY, which for the past 13 years has been fostering healthy lives through nutrition education, physical activities, and wellness initiatives (Flipany.org 2019). They offer cooking classes through a "Train the trainer" methodology and provide healthy food to local schools through the after school and summer meal programs.

There are also some national organizations that have made an impact on Greater Miami's community nutrition resilience. For example, Wholesome Wave, which aims to empower under-served consumers to make better food choices by increasing affordable access to healthy produce. Their two flagship programs are the doubling of the value of SNAP benefits when spent on fruits and vegetables, and their work with doctors to prescribe produce. They have a network of 1400+ farmers markets and grocery stores in 49 states (Wholesomewave.org 2019). In Greater Miami, they partnered with the retailer Target to offer vegetable prescription programs to economically disadvantaged Miami residents.

There are numerous other organizations that work with providing hunger relief during extreme weather events, and it would be impossible to discuss all. These organizations include Make the Homeless Smile, Bridge to Hope, New Florida Majority, WeCount, Family Action Network Movement (FANM), Community Justice Project, Florida Immigrant Coalition (FLIC), the Miami Workers Center (MWC), MH Action, PowerU, Sant La Haitian Neighborhood Center, SEIU Florida and the Miami Climate Alliance (MCA). For example, Valencia Gunder and Vanessa Tinsley run two nonprofits that provide food to residents after severe weather events. After Hurricane Irma, Valencia and Make the Homeless Smile provided 20,000+ meals in Miami's urban neighborhoods, while Vanessa and Bridge to Hope gave out 20,000 bags of groceries in Homestead. They have started using grant dollars from the hurricane relief funds of the Miami Foundation to stock up on supplies and get community-led emergency response teams in place (WLRN.org 2018).

The MCA is another noteworthy organization. Its mission is to build urgency, power, and cohesion by activating the Miami community through strategic actions that recognize climate change as a threat multiplier to all forms of justice, especially for low-income frontline communities, and to create a model for just, equitable, and resilient communities in the face of climate change by achieving substantive wins that will increase the well-being of Miami residents in both the present and future (Miamiclimatealliance.org 2019). The Alliance includes more than 90 member organizations, including some discussed above and below.

It is expected that the work of these organizations has led to notable changes in fruit/vegetable consumption in low-income communities in Miami and have helped low-income communities have and maintain access to healthful foods following natural disasters like Hurricane Irma. However, data on the impacts of these programs are lacking.

When asked about key nonprofits working in the climate adaptation space, interviewees repeatedly mentioned two: the CLEO Institute and Catalyst. According to Torriente (2018), these organizations influence local government and have been advocating to focus on social issues.

The CLEO Institute has a great reputation to have a non-partisan focus on climate-based science. According to Yoca Arditi-Rocha (2018), Executive Director of the CLEO Institute, among the many climate change mitigation and resilience-related programs they lead are the Climate Speakers Network, which empower citizens to become speakers and change agents on the topic. Further, their Empowering Resilient Women program looks at educating women, as the head of the household, to be better prepared for hurricanes and provide them with leadership skills to advocate for other causes in their community. CLEO's Empowering Youth program takes kids on field trips to look at pumps, raised streets and other resilience efforts and invites them to develop a resilience for their own neighborhood. Further, they lead mayoral tables as opportunities for mayors to learn from each other, as well as trainings for staff members of municipalities.

Catalyst is another important nonprofit working in the space. Established by now-Commissioner Daniella Cava Levine in 1996, they have traditionally been focusing on human services work. In 2015, with a grant by the Kresge Foundation, they started climate-related work, and, for example, organized the first climate march in Miami. Catalyst still mainly does work related to economic stability of communities. For example, they offer financial coaching to under-privileged neighborhoods. On climate resilience, they provide leadership training for communities, teaching how to organize communities around issues of social justice and climate resilience in Miami-Dade, ultimately aiming to provide opportunities for people to participate on government level.

3.6.2 Community Foundations

In Greater Miami which is less than 120 years old and thus still lacks systemic, government-led leadership, foundations have traditionally played an important role in progressing on critical challenges in the region.

There are some that have started working on community nutrition-related programs. The Rockefeller Foundation with its 100RC initiative has previously been mentioned as one of the key catalysts of a new focus on resilience in Greater Miami, particularly GM&B. It also offers support around food security, which GM&B is currently not taking advantage of.

Further, The Miami Foundation which was established in 1967 aims at bridging profit and nonprofit sectors for common good of community to advance quality of life in Miami (Miamifoundation.org 2019). With a focus on SLR adaptation and climate literacy, they support programs that increase public green spaces to reduce the heat island effect and control flooding and bring communities close to nature. As previously discussed, the Health Foundation of South Florida supports the Live Healthy Miami Gardens and Live Healthy Little Haiti programs. In addition, the Kresge Foundation, through their support of the Southeast Florida Regional Climate Change Compact, has been instrumental in fostering collaboration and action on a regional level.

In addition, there appears to be a new wave of philanthropists in Miami that support climate change issues for the well-being of communities. Examples are Social Venture Partners, a group of Venture Philanthropists (Socialventurepartners.org 2019), which supports non-profit as well as for-profit organizations in Greater Miami that have a positive impact on critical challenges affecting communities, including nutrition.

3.6.3 Academia

The local academic institutions in Greater Miami do not appear to be focusing on community nutrition resilience directly. However, there appear to be some opportunities for a future integration of the topic.

At University of Miami (UM), its Office of Civic and Community Engagement (CCE) fosters university-community collaborations by engaging the university's academic resources in the enrichment of civic and community life in our local, national, and global communities (University of Miami 2019a). CCE created the Miami Housing Solutions Lab to provide resources, tools, and data related to affordable housing and community development in the Miami metro area. This platform provides community groups, planners, policy makers, and affordable housing developers with information on local housing needs as well as housing policies that prevent displacement and promote affordable housing (Miami Housing Solutions Lab 2019). According to Marisa Hightower, Associate Director at the CCE, at this time, UM is not focused on food security/food deserts. Although there is discussion within the university to look into GIS project around these topics, it still is in nascent stages. UM also organizes an annual food day and has their own food garden which is available to the public (University of Miami 2019b). Notably, a former student created a story map about community gardens in Greater Miami, which however wasn't finished (Vildosola 2019).

Notably, Florida International University (FIU) runs StartUP FIU Food, an incubator that serves local food and beverage business owners that want to grow and scale their companies. It is unclear how much this program prepares food businesses for chronic stressors and acute shocks.

3.6.4 Private Sector

As previously discussed, local businesses in Greater Miami are represented by the Chambers of Commerce and the Beacon Council, which both have a working group on resilience, but do limited work.

Overall, activities in the field of community nutrition resilience by the private sector seem to focus on donating food to local food banks. Nationwide and over the years, General Mills has given US$57 million worth of food to food banks in the USA and internationally; Campbell's has donated over US$50 million in food to support organizations like

Feeding America, as well as put healthy food into 40 corner stores in areas classified as food deserts; and PepsiCo. has set the lofty goal of eliminating food deserts entirely (MindBodyGreen.com 2018).

Walmart is an example of a business with a strong focus on resilience of its supply chain, heightened by hurricane Katrina. Walmart has an Emergency Operations Center (EOC) that monitors weather and other disruptions to business operations. The EOC serves as a hub to engage operations teams throughout the business and ensure that stores are prepared for a disruption. Walmart is able to pre-position supplies in emergency warehouses strategically located throughout the country to ensure ready access after a disaster. Walmart works to restore store operations as soon as possible and can get most stores up and running within 24 hours, depending on the severity of the disaster and its impact on transportation routes. Walmart has leveraged its logistics expertise to support community recovery after disasters, including Hurricane Katrina, Superstorm Sandy and the 2016 Louisiana floods. In 2015, Walmart also launched a pilot project focused on building community capacity to respond to a disaster, working with local government and nonprofit organizations. BRIDGe Corps (Building Resiliency in Disaster-prone Geographies) New Orleans was a short-term volunteer initiative that matched senior level Walmart employees with the City of New Orleans to improve emergency management operations related to emergency food supply storage (e.g., water and Meals Ready to Eat) and warehouse logistics (Zeuli and Nijhuis 2017).

Locally, there are some other examples of businesses taking note, or action. As previously mentioned, Target is collaborating with several organizations such as Wholesome Wave and FLIPANY to improve access to healthy, nutritious food for local communities. Another example is Baptist Health System. In their Homestead hospital, they have been dedicating land around their facility to their "Grow2Heal" program for the past 4 years. Led by Thi Squire, a community garden was created which now serves the hospital canteen as well as the local community. In addition, it offers education programs for schools and community groups, teaching cooking skills and the basic principles of healthy eating. The program is still expanding the acreage and is piloting a vegetable prescription program (Squire 2018).

Overall, there is still a huge opportunity to engage the private sector more and collaborate on the issue of community nutrition resilience that affects their local workforce.

3.6.5 Media

The media can play an important role in fostering community engagement and activism. And although community nutrition resilience has not been picked up directly by the local media, climate change resilience has, particularly through a collaboration of local media outlets. The Invading Sea is a collaboration by the editorial boards of the South Florida Sun Sentinel, Miami Herald and Palm Beach Post—with reporting by WLRN Public Media—to address the threat South Florida faces from SLR. They aim to raise awareness, amplify the voice of our region, and create a call to action that can't be ignored (The Invading Sea 2019).

The daily newsletter New Tropic issued by WhereBy.Us Enterprises is another publication that is vocal about the USA's, Florida's, and Greater Miami's climate change mitigation efforts, as well as about their response to threats from climate change (The New Tropic 2019). The Miami-based organization curates content that informs its subscribers about the latest developments in environmental legislation and activities in Miami's communities.

3.6.6 Civil Sector

On a civic level, there has been an increase in engagement in Greater Miami in recent years related to climate issues although activity is still lower than in other cities in the USA, even though those might be less at risk for climate change impacts. A University of Notre Dame (2019) index which has ranked 270 U.S. cities on more than 40 climate metrics including climate readiness as driven by, for example, civic engagement and climate change awareness, ranks Miami at 202, Hialeah at 264, Fort Lauderdale at 61, and West Palm Beach at 180. Interestingly, Pembroke Pines which appears not to be doing much, ranks 102.

In Miami, activities are punctual. For example, the March for Science Miami, founded in 2017, drew over 4000 people in that year (Sciencemarchmiami.org 2019). In 2019, the annual "100 Great Ideas" project spearheaded by Miami-based social impact accelerator Radical Partners was dedicated to the topic climate resilience and sustainability, inviting local residents to brainstorm over 5 days on solutions to both. The top three themes that emerged were energy-efficient buildings and homes, transportation and water. Food systems came fifth, and ideas included reducing meat consumption, growing edible plants in public spaces, introduce municipal composting and redistribute unused food to avoid food waste (Radical Partners 2019). The CLEO Institute's 2019 "Empowering Capable Climate Communicators Symposium" drew over 200 participants from different parts of the community to learn and discuss the causes and impacts of climate change (The CLEO Institute 2019).

When aiming to design solutions to address community nutrition resilience challenges in Greater Miami—ideally some that complement what has already been done—it seems useful to look at what other cities have been doing to address the same challenge. The following chapter is hence going to discuss selected cases that have been mentioned by interview partners or have been identified in the research for this book.

REFERENCES

100ResilientCities.org. (2018). *Frequently asked questions (FAQ) about 100 Resilient Cities* [online]. Available at: http://www.100resilientcities.org/100RC-FAQ/#/-_/. Accessed 29 January 2018.

Beacon Council. (2019). *One community one goal: Preparing Miami-Dade County for long-term economic growth* [online]. Available at: https://www.beaconcouncil.com/ocog/. Accessed 20 February 2019.

Broward.org. (2019). *Broward County land use plan* [online]. Available at: http://www.broward.org/BrowardNext/Pages/broward-county-land-use-plan.aspx. Accessed 14 January 2019.

Business Insider. (2018). *Miami is racing against time to keep up with sea-level rise* [online]. Available at: https://www.businessinsider.com/miami-floods-sea-level-rise-solutions-2018-4. Accessed 10 October 2018.

Ces.fau.edu. (2014). *Risk, resilience and sustainability: A case study of Fort Lauderdale* [online]. Available at: http://www.ces.fau.edu/publications/pdfs/fort-lauderdale-case-studyv3.pdf. Accessed 16 October 2018.

Citizens' Climate Lobby. (2018). *What is the Climate Solutions Caucus?* [online]. Available at: https://citizensclimatelobby.org/climate-solutions-caucus/. Accessed 23 December 2018.

Commonthreads.org. (2019). *Home* [online]. Available at: http://www.commonthreads.org/. Accessed 12 February 2019.

Farmshare.org. (2019). *About us* [online]. Available at: http://farmshare.org/about-us/. Accessed 2 March 2019.

Feeding South Florida. (2018). *South Florida continues to face hunger challenges* [online]. Available at: https://feedingsouthflorida.org/south-florida-continues-to-face-hunger-challenges/. Accessed 20 February 2019.

Flipany.org. (2019). *What we do* [online]. Available at: http://flipany.org/about-flipany/what-we-do. Accessed 10 February 2019.

Florida Phoenix. (2019). *How much is climate change costing us? A bill in the Legislature would start an official accounting* [online]. Available at: https://www.floridaphoenix.com/2019/03/29/how-much-is-climate-change-costing-us-a-bill-in-the-legislature-would-start-an-official-accounting/. Accessed 13 May 2019.

Foodrescue.net. (2019). *K-12 Food Rescue* [online]. Available at: https://www.foodrescue.net/. Accessed 5 March 2019.

Fortlauderdale.gov. (2018). *Vision plan: Fast forward Fort Lauderdale: Our city, our vision 2035* [online]. Available at: https://www.fortlauderdale.gov/departments/city-manager-s-office/structural-innovation-division/vision-plan. Accessed 16 October 2018.

Gimenez, C. (2018, October). Keynote given at the 10th Annual Southeast Florida Regional Climate Leadership Summit, Miami Beach, FL.

Healthymiamidade.org. (2019). *Home* [online]. Available at: https://www.healthymiamidade.org/. Accessed 20 February 2019.

Houston, M. (2018, October 17). Personal interview.

ICMA.org. (2018). *Esri smart communities case study series: Ft. Lauderdale: Developing a resilient, smart city* [online]. Available at: https://icma.org/sites/default/files/308254_16-276%20Esri%20Case%20Study%20Ft%20Lauderdale-web.pdf. Accessed 15 October 2018.

Jurado, J. (2018, October 17). Personal interview.

LaPradd, C. (2018, November 2). Personal interview.

Mdc.maps.arcgis.com. (2018). *What is Miami-Dade County doing about sea level rise?* [online]. Available at: https://mdc.maps.arcgis.com/apps/Shortlist/index.html?appid=51003eca3778442ca5b8bc8c0868920a&mc_cid=d589361308&mc_eid=5e441ffbf9. Accessed 20 September 2018.

Miami Herald. (2017). *Miami gets $200 million to spend on sea rise as voters pass Miami Forever Bond* [online]. Available at: http://www.miamiherald.com/news/politics-government/election/article183336291.html. Accessed 29 January 2018.

Miami Herald. (2018a). *Curbelo considering 2020 Miami-Dade mayoral bid* [online]. Available at: https://www.miamiherald.com/news/politics-government/article222319520.html. Accessed 30 November 2018.

Miami Herald. (2018b). *How will sea level rise affect your home? Miami is creating a tool that will show you* [online]. Available at: http://www.miamiherald.com/news/local/community/miami-dade/article209710464.html. Accessed 4 May 2018.

Miami Herald. (2018c). *Florida leads nation in property at risk from climate change* [online]. Available at: https://www.miamiherald.com/news/local/environment/article29029159.html. Accessed 12 April 2019.

Miami Housing Solutions Lab. (2019). *Miami housing solutions lab* [online]. Available at: http://cdn.miami.edu/wda/cce/Documents/Miami-Housing-Solutions-Lab/index.html. Accessed 19 January 2019.

Miamibeachfl.gov. (2018). *Sustainability plan: Energy economic zone plan* [online]. Available at: https://www.miamibeachfl.gov/wp-content/uploads/2017/12/City-of-Miami-Beach-Sustainabilty-Plan_FINAL.pdf. Accessed 22 October 2018.

Miamiclimatealliance.org. (2019). *Home* [online]. Available at: http://miamiclimatealliance.org/. Accessed 10 February 2019.

Miamifoundation.org. (2019). *About us: Advancing quality of life in Greater Miami* [online]. Available at: https://miamifoundation.org/about/. Accessed 10 January 2019.

MindBodyGreen.com. (2018). *40 million Americans don't have access to enough food: These companies are trying to change that* [online]. Available at: https://www.mindbodygreen.com/articles/naked-juice-wholesome-wave-pepsi-food-access-initiatives. Accessed 12 February 2019.

Myboca.us. (2018). *Media release: The city of Boca Raton hires first sustainability manager* [online]. Available at: https://www.myboca.us/DocumentCenter/View/17314/Media-Release---City-of-Boca-Raton-Hires-First-Sustainability-Manager-05312018-PDF. Accessed 16 October 2018.

Palm Beach County. (2015, October). *Hunger relief plan Palm Beach County* [online]. Available at: http://discover.pbcgov.org/communityservices/humanservices/PDF/News/Palm_Beach_FRAC_100515_Edition-v2.pdf. Accessed 25 January 2019.

Radical Partners. (2019). *100 great ideas* [online]. Available at: https://static1.squarespace.com/static/56526d51e4b083936d517729/t/5ca4d099eb3931713a53a29b/1554305178677/100+Great+Ideas+Climate+Resilience+%26+Sustainability+Final+Report.pdf. Accessed 10 May 2019.

Resilient305. (2019). *Resilient Greater Miami & the beaches*. Miami, FL.

Resilient305.org. (2018). *Resilient Greater Miami & the beaches: Preliminary resilience assessment #Resilient305* [online]. Available at: http://resilient305.com/assets/pdf/170905_GM&B%20PRA_v01-2.pdf. Accessed 29 January 2018.

Sciencemarchmiami.org. (2019). *Who we are* [online]. Available at: https://www.sciencemarchmiami.org/. Accessed 10 May 2019.

Socialventurepartners.org. (2019). *Social Venture Partners Miami: A global venture philanthropy model arrives* [online]. Available at: http://www.socialventurepartners.org/wp-content/uploads/2017/01/Miami-Overview-Interest-Form.pdf. Accessed 10 February 2019.

Southeastfloridaclimatecompact.org. (2017). *Advancing resilience solutions through regional action* [online]. Available at: http://www.southeastfloridaclimatecompact.org/. Accessed 18 December 2017.

Southeastfloridaclimatecompact.org. (2019a). *Public health* [online]. Available at: http://www.southeastfloridaclimatecompact.org/recommendation-category/ph/. Accessed 5 January 2019.

Southeastfloridaclimatecompact.org. (2019b). *Agriculture* [online]. Available at: http://www.southeastfloridaclimatecompact.org/recommendation-category/ag/. Accessed 5 January 2019.

Squire, T. (2018, November 19). Personal interview.

The CLEO Institute. (2019). *Empowering capable climate communicators symposium* [online]. Available at: https://www.cleoinstitute.org/symposium-2019. Accessed 10 May 2019.

The Invading Sea. (2019). *About us* [online]. Available at: https://www.theinvadingsea.com/about-us/. Accessed 10 February 2019.

The Hill. (2018). *Poll: Record number of Americans believe in man-made climate change* [online]. Available at: https://thehill.com/policy/energy-environment/396487-poll-record-number-of-americans-believe-in-man-made-climate-change. Accessed 2 February 2019.

The New Tropic. (2019). *About us* [online]. Available at: https://thenewtropic.com/about/. Accessed 10 May 2019.

Thepatchgarden.com. (2019). *About* [online]. Available at: http://thepatchgarden.com/about. Accessed 10 February 2019.

Torriente, S. (2018, October 24). Personal interview.

UNFCC.int. (2019). *What is the Paris Agreement?* [online]. Available at: https://unfccc.int/process-and-meetings/the-paris-agreement/what-is-the-paris-agreement. Accessed 3 January 2019.

University of Miami. (2019a). *Welcome* [online]. Available at: http://www.miami.edu/civic. Accessed 10 February 2019.

University of Miami. (2019b). *Sustainability* [online]. Available at: https://greenu.miami.edu/topics/food-and-well-being/community-gardens/index.html. Accessed 10 February 2019.

University of Notre Dame. (2019). *Urban adaptation assessment: Better data for planning for your city's future* [online]. Available at: https://gain-uaa.nd.edu/. Accessed 23 May 2019.

Urbangreenworks.org. (2019). *Urban GreenWorks, restoring the economic, physical and social health of under-served communities* [online]. Available at: https://www.urbangreenworks.org/. Accessed 10 February 2019.

Urbanland.uli.org. (2018). *Living with rising sea levels: Miami Beach's plans for resilience* [online]. Available at: https://urbanland.uli.org/sustainability/living-rising-sea-levels-miami-beachs-plans-resilience/. Accessed 22 October 2018.

US Army Corps of Engineers. (2019). *Miami-Dade Back Bay coastal storm risk management feasibility study* [online]. Available at: https://www.saj.usace.army.mil/MiamiDadeBackBayCSRMFeasibilityStudy/. Accessed 2 February 2019.

Vildosola, D. (2019). *Community gardening in the first step to sustainable living* [online]. Available at: https://umiami.maps.arcgis.com/apps/MapJournal/index.html?appid=88c1fbb1c06d441298a4ac2fb2574333. Accessed 10 February 2019.

Wholesomewave.org. (2019). *Changing the world through food* [online]. Available at: https://www.wholesomewave.org/. Accessed 12 February 2019.

Wlrn.org. (2018). *Community groups begin work on hurricane plans for low-income neighborhoods in Miami-Dade, Broward* [online]. Available at: http://wlrn.org/post/community-groups-begin-work-hurricane-plans-low-income-neighborhoods-miami-dade-broward. Accessed 7 May 2018.

WPB.org. (2018). *Sustainability overview* [online]. Available at: http://wpb.org/Departments/Sustainability/Overview. Accessed 16 October 2018.

Wyncstations.org. (2018). *The Takeaway: In face of looming challenges, building a resilient Miami* [online]. Available at: https://www.wnycstudios.org/story/mayor-francis-suarez-his-first-term-office?mc_cid=a44813592e&mc_eid=5e441ffbf9. Accessed 20 September 2018.

Zeuli, K., & Nijhuis, A. (2017). *The resilience of America's urban food systems: Evidence from five cities* [ebook]. Roxbury, MA: ICIC. Available at: http://icic.org/wp-content/uploads/2017/01/Rockefeller_ResilientFoodSystems_FINAL_post.pdf?x96880. Accessed 18 December 2017.

CHAPTER 4

Designing Nutrition Resilient Communities: Learnings from Other Cities

Abstract To determine recommendations regarding how to better ensure community nutrition resilience in Greater Miami, this chapter looks at examples of other cities that have addressed the topic in their policy and planning efforts. It examines Boston's similar challenges, New York City's integration of food systems in the city's resilience plan as well as the transparent measurement and reporting of progress, and Baltimore's integrated food policy efforts. Also discussed are Orlando's leadership and integrated planning and programs and Los Angeles' food policy council. The chapter also summarizes learnings from other cities, for example, related to their approach to the cultural change that is required to address community nutrition resilience. Finally, food waste and recovery is examined as an area that is gaining momentum.

Keywords Boston · New York City · Baltimore · Orlando · Los Angeles · Food waste and recovery

As the previous chapter has demonstrated, in Miami-Dade County there is very little to no focus on food systems and community nutrition resilience, both from a policy and planning perspective, as well as related to any coordinated efforts from organizations in the public and private sector. Although some work is being done by nonprofit organizations to

© The Author(s) 2020
F. Alesso-Bendisch, *Community Nutrition Resilience in Greater Miami*, Palgrave Studies in Climate Resilient Societies, https://doi.org/10.1007/978-3-030-27451-1_4

relieve hunger and increase community resilience, it does not tie into any broader strategy.

Broward County, with their updated land use and comprehensive plan elements on food policy, has provided a basis for local government action. Punctual efforts particularly related to food recovery and with support from the school district are noteworthy. Although the ultimate goal of the program has been GHG emission reduction, it already has made an impact on several families in the county.

Palm Beach County's Hunger Relief Plan addresses food security as well as health literacy. However, systematic work on eliminating vulnerabilities in the local food systems, as well as addressing community nutrition resilience in the face of chronic stresses and acute shocks is missing. Neither of the three counties has a food policy council to devise and guide policy and programs.

This goal of this chapter is to provide learnings from case studies of other cities in the USA which operate in similar political, economic, environmental, and social circumstances as the cities in Greater Miami. Further, it draws on examples of other international cities in the 100RC, as well as on other cities mentioned during the interviews for this book and identified through further research.

4.1 Boston

Boston is part of the 100RC network and although it has not identified food security as one of their main challenges, it has been leading the way on food system and community nutrition resilience.

Despite being an important cultural and economic center, rich with US history, certain neighborhoods, especially those with minority communities, have not prospered like the rest of the city. Like in Greater Miami, lack of affordable housing, fewer educational opportunities, and less preparation for good jobs threaten to divide the city further along racial and economic lines. The city has launched new programs to increase access to jobs, including training and related services, in order to help disadvantaged communities, move into growing industries and promote equity throughout the city (100ResilientCities.org. 2018d).

In many ways, Boston faces similar challenges to its resilience as Miami. It particularly needs to develop plans to respond to flooding and the impacts of sea level rise. Several essential civic and transportation hubs, including Logan International Airport, are located in flood-sensitive zones.

Resilience efforts in Boston started in 2014, when it became the first city of its size in the USA to study the impact of a natural disaster on its food system. It was motivated by the near miss of Superstorm Sandy, which destroyed food system assets in wide parts of the country. As the newly elected Mayor Walsh wrote in support of the study, *"Boston was lucky to avoid the worst of Sandy, but with climate change we can expect a rise in sea levels and more extreme weather events in the future. We must better prepare our food system to be resilient after disruptions like hurricanes, floods, blizzards and other natural disasters"* (Zeuli and Nijhuis 2017: 51).

The impact study was commissioned by the City of Boston's Office of Food Initiatives (OFI), Office of Emergency Management, Office of Environment, Energy and Open Space, and the Transportation Department in 2014. The study culminated in a set of 17 recommendations and an implementation roadmap to address gaps in information and provide direction to the City for strengthening Boston's food system to ensure it can quickly return to normal conditions following a natural disaster (Zeuli and Nijhuis 2017).

The research found that food availability and food access could be significantly compromised in the event of a natural disaster in Boston. The study helped inform climate resilience priorities for the Massachusetts Food Policy Council, a 17-member entity comprised of state agency, legislative, and industry representatives. The Council's new Massachusetts Local Food Action Plan, released in 2015, includes a number of recommendations to improve the resilience of the state's food system. It covered (ICIC.org. 2015a):

1. Greater public–private food system coordination, within Boston and the region,
2. Investment in critical food system infrastructure, including buildings and roads,
3. More national chain grocery stores in low-income neighborhoods,
4. Robust resilience plans for small grocery stores and corner stores, and
5. Expanded capacity and increased efficiency of Boston's food safety net.

To implement these recommendations, Boston developed an implementation roadmap, charging critical agencies or organizations with steps for implementation. For example, a newly established Office of Food Initiatives (OFI), reporting to the Chief of Health and Human Services, established a multi-stakeholder food system resilience committee to address, for example, the protection of local food infrastructure, businesses, and organizations. Further, Boston's Redevelopment Authority and Office of Business Development started strengthening the resilience of the City's food system organizations and private-sector assets. And the Greater Boston Food Bank began developing strategies to improve efficiency and expand capacity (ICIC.org. 2015b).

Shortly after the release of the report, Boston was selected to join the 100RC network and hired its first CRO. The CRO had been a member of the study's steering committee and has been working on incorporating recommendations from the food system resilience study into the City's resilience strategy.

The study also helped inform the priorities of the Metro Boston Climate Preparedness Taskforce, a group of 14 Greater Boston municipalities working to coordinate regional, cross-government action to prepare for the effects of climate change. It has identified strengthening the resilience of fresh food distribution centers as one of its three 2016 priorities and is conducting additional research on the flood risks facing the region's food distribution infrastructure.

Boston is working toward implementing some of the report's recommendations, although internal reorganizations pose a challenge. For example, the Director of the OFI was tasked with establishing a food system resilience committee that would include representatives from public- and private-sector food organizations. This new committee would have to be responsible for coordinating food system resilience planning. During the study, however, the OFI was restructured with new leadership and staff. As a result, some of the food system resilience planning has lost momentum and now competes with their core focus on nutrition, food access, urban farming, and food truck licensing. The OFI however continues to be committed to support research on food access in neighborhoods with high rates of poverty and food insecurity (Zeuli and Nijhuis 2017).

The case of Boston demonstrates that even with widespread support within the City, including from the Mayor's Office, resilience planning of any type is complicated and it takes a long time to build sustainable coalitions that can ultimately implement the plans. In addition, long-term planning competes with the numerous urgent, immediate issues facing public agencies and the Mayor's Office. And, finally, food system resilience planning involves many agencies and does not fit neatly into a single agency, meaning that it always competes with core agency priorities.

Boston's approach and challenges are relevant when looking at strengthening Greater Miami's community nutrition resilience in the wake of chronic stresses and acute shocks. In particular, counties and cities in Greater Miami can take note of the importance of starting out with a comprehensive evaluation of vulnerabilities related to the local food system. This should be followed by an integration of food system resilience into a city's resilience planning and the development of actionable implementation steps. Also key is having a clear allocation of responsibility for implementation, as well as the need for multi-stakeholder collaboration.

4.2 New York City

New York City, which is also part of 100RC, provides another example related to good governance and processes and also offers interesting insights into communication.

Just like in Boston, New York City's food priorities are directed by the Office of the Director of Food Policy. The office coordinates multiple City agencies and offices that work on food programs or policies, as well as partners with the many advocates and nonprofit organizations working in food. The office also convenes the Food Policy Task Force, comprised of representatives from multiple City agencies and the City Council (NYC Food Policy 2019b).

Improving the resilience of the city's food supply is a key part of New York City's resilience plan, One New York: The Plan for A Strong and Just City. The plan includes an initiative to invest in making the city's fresh food distribution center (Hunts Point) more resilient to better prepare it for power outages, coastal flooding, job losses, and other disruptions from extreme weather events (Zeuli and Nijhuis 2017).

As part of this plan, New York City has been launching a number of initiatives specifically focused on reducing household and commercial food waste. The City has set a Zero Waste goal by 2030, which would reduce the amount of all waste, including food waste, by 90%.

The New York City Organics Collection Program, which is run by the Department of Sanitation in select neighborhoods, will be expanded citywide to collect food scraps and other organic material from households to be used for composting or conversion to biogas energy. To reduce commercial food waste, Mayor Bill de Blasio launched the Mayor's Food Waste Challenge, which encourages New York City restaurants to reduce food waste to landfills by 50% (Zeuli and Nijhuis 2017).

The city also commissioned a study on the resilience of their food supply chains in 2016, driven by its Economic Development Corporation (NYCEDC) and the Mayor's Office of Recovery and Resiliency. Dubbed Five Borough Food Flow, it analyzed the resilience of the New York City region's food distribution system.

Also notably, New York City is measuring its progress on food security. Already in 2011, the New York City Council established reporting requirements for a variety of City agency initiatives related to food Local Law 52 of 2011. The Food Metrics Report provides a snapshot of data from those programs as well as trends over time. The report has expanded every year to include the broad range of programs and initiatives that the City is doing to address food insecurity; improve City food procurement and food service, increase healthy food access and awareness, and support a more sustainable and just food system (NYC Food Policy 2019c). The city publishes information about its food programs, metrics, and policies on a website to inform its residents (NYC Food Policy 2019a).

In terms of lessons for Greater Miami, particularly New York City's clear allocation of responsibility to a Director of Food Policy, already seen in the case of Boston, should be noted. Further, there appears to be a strong integration of food system resilience in the city's resilience plan. Progress is measured and reported transparently. Finally, New York City's efforts to reduce food waste can be considered best practice.

4.3 Baltimore

Though not a part of the 100RC network, the City of Baltimore's work to strengthen the resilience of its food system seems relevant. Focused primarily on food environment disparities, the Department of Planning and the Johns Hopkins Center for a Livable Future have been collaborating to examine the Baltimore food environment through research, analysis, and mapping in order to inform the work of the City's Baltimore

Food Policy Initiative (BFPI). BFPI is a collaboration between the Department of Planning, Office of Sustainability, Baltimore City Health Department and Baltimore Development Corporation (Baltimorecity. gov. 2019).

As part of their food system efforts, Baltimore notably uses the term "Healthy Food Priority Areas" instead of "food deserts." Multi-stakeholder consultations revealed that the term "food desert" is often met with critique or disapproval. "Food desert" suggests there is no food, when in actuality in urban environments there is an imbalance between healthy and unhealthy food. Additionally, the term puts the whole area in a liability framework and does not acknowledge the amazing grassroots work occurring on the ground to fill the gaps. As a result, the term "Food Deserts" has been changed to "Healthy Food Priority Areas."

In Baltimore, almost 24% of residents live in Healthy Food Priority Areas. And just like in Greater Miami, certain groups of residents are affected at disproportional rates. More than 31% of residents who live in Priority Areas are Black, compared to only 8.9% of White residents. Children are with 28% the most likely of any age-group to live in Priority Areas (Baltimore City 2018).

Also noteworthy is that the City issues so-called Food Environment Briefs, which along with the input of Resident Food Equity Advisors drive the city's comprehensive eight-point Healthy Food Environment Strategy. These briefs provide a snapshot of the impact of Healthy Food Priority Areas and an analysis of food retail, nutrition assistance, and urban agriculture from a citywide perspective. Further in 2018, comprehensive briefings for each of the fourteen City Council Districts and six State Legislative Districts were created. This information helps policy makers understand what the food system looks like in their districts as well as citywide. Resident Food Equity Advisors are available to support the efforts in their districts.

Baltimore's Healthy Food Environment Strategy comprehensively addresses the different steps of the food system. It has the following goals (Baltimorecity.gov. 2019):

1. Support resident-driven processes to guide equitable food policy, priorities and resources;
2. Improve small grocery, corner and convenience stores;
3. Retain and attract supermarkets;

4. Increase the ability of the public markets to anchor the healthy food environment;
5. Implement supply chain solutions that support healthy food distribution and small businesses;
6. Maximize the impact of nutrition assistance and meal programs;
7. Support urban agriculture, emphasizing historically disenfranchised populations and geographies; and
8. Address transportation gaps that impact food access.

From looking at the case of Boston, several points that seem to be working well can be noted. First, the city's Food Policy Initiative, a collaboration between four city departments guiding policy. Secondly, the initial assessment of the food environment that has been conducted with John's Hopkins as well as the regular and transparent sharing of data through the briefs. Further, the used terminology of "Healthy Food Priority Areas" instead of food deserts seems useful. Finally, Baltimore's eight-goal, comprehensive strategy addressing all steps of the food system.

4.4 Orlando

Orlando's efforts on food systems and food security have been frequently cited in the interviews as a best practice example in South Florida. The work is led by Chris Castro, Director of Sustainability and Resilience in Orlando, and under the leadership of Mayor Buddy Dyer. Among the things that work well in Orlando, the political support and leadership have been mentioned, further to its inclusive, multi-sectoral approach to sustainability planning.

Orlando's Office of Sustainability and Resilience was established in 2007 by Mayor Dyer. The City's sustainability efforts are summarized in Green Works Orlando, which covers the areas of people, planet, and prosperity. According to Mayor Dyer, Orlando's vision is to "transform Orlando into one of the most environmentally-friendly, economically and socially vibrant communities in the nation" (City of Orlando 2019: 1).

Historically, Orlando started out with a focus on the City's footprint, and only afterward looked at its systems. According to Castro (2018), within the first five years, Orlando focused on improving their own environmental footprint. This seems to be a common approach, since city leaders need to demonstrate leadership by example. Since Orlando has their own municipal utility, it was arguably able to move quicker toward their emission reduction goals than Greater Miami, which relies on Florida Power & Light (FPL) to enable these efforts.

In 2013, Orlando then started to look outwards. Named Orlando's Green Work Community Action Plan, their community action plan was shaped in collaboration with partners of government sectors, academia, and other sectors. It focuses on seven pillars: clean energy, green buildings, local food, livability, solid waste, transportation, and water. These key focus areas were chosen by the community. Achievements in these areas are published online on the City of Orlando Web site (City of Orlando 2019).

Within the area of local food, the goal has been to make Orlando a leading food destination.

This includes making the local food system including its assets more resilient against climate change, and to enhance food policy, production and distribution. Among the many areas, Orlando is addressing underneath this pillar are farm to table restaurants, farmers' markets, community gardens and farms, community kitchens, and front yard farming.

Equally important, Orlando aims to provide every resident with access to affordable and healthy food. This should be achieved by means of either a grocery store, a community farm, or garden, or any other access point being no further than 1.5 mile from every resident. This is enabled by increasing local food assets (Castro speaks of a growth by 10 times) and results have been impressive. The number of farmers markets, for example, has increased from six to now 140 in Orlando perimeters. Further, about 50 homeowners are now participating in front yard farming.

A key prerequisite and as Castro calls it "foundational step" is a robust food policy that supports local urban agriculture. That is why Orlando established Good Food Orlando, its multi-stakeholder food policy council. Led by the City, participation has rotated throughout the years of its existence, but has included partners from food banks, the media, Government, UF IFAS, the Department of Health and local food entrepreneurs and practitioners. Sub-working groups have been established that work on topics covering different steps of the food systems, from production, through processing, distribution and waste/recovery. There are goals and targets for each of the groups. As education is one of the main goals of the council, the monthly meetings are being held at the site of different participating organizations, with the representative of this organization chairing that meeting. According to Castro, if any city is looking to enhance its local food system, it has to establish a food policy council that has this mission.

Further to the council, Castro (2018) argues that the food system has to be considered in local planning. In the case of Orlando, the East Florida Regional Planning Council conducted a number of studies of the local food system and mapped it out. Based on this effort, they developed a food system planning roadmap for Central Florida. According to Castro, it is critical for any city to consider the food systems in their planning efforts and thus should become an initial goal for work in Greater Miami.

Orlando, as a city in relatively close proximity to Greater Miami, provides many insights about how to implement a successful focus on food system and nutrition resilience. Some of these have been mentioned earlier for other case studies. However, in addition to Orlando's integrated and multi-stakeholder approach to planning and implementation, driven by strong leadership from both Mayor Buddy as well as Chris Castro, the learnings related to the Food Policy Council must be noted.

4.5 Los Angeles

Just like the cities discussed earlier, Los Angeles' work on community nutrition resilience also started with a study of the resilience of their food supply chains. In 2015, Los Angeles' Emergency Management Department (EMD) received technical assistance planning support from the FEMA National Integration Center to complete a high-level, food supply chain resilience study. It analyzed the impact of a 7.8 magnitude earthquake on large grocery store supply chains.

Particularly noteworthy is Los Angeles' Good Food Policy Council. Founded in 2011, it serves as backbone organization for a network of over 400 organizations and agencies working for healthy, sustainable, and fair food. This collaboration of so many actors is made possible through the application of the collective impact model.

Collective impact refers to the commitment of a group of important actors from different sectors to a common agenda for solving a specific social problem (Stanford Social Innovation Review 2011). In the case of the multidisciplinary work of the Los Angeles Good Food Policy Council, it is used to make transformative change in the six impact areas food security, food equity and access, street food, regenerative and urban agriculture, food waste and good food purchasing policy (Los Angeles

Food Policy Council 2019). It thus harnesses the power of many to multiply the results of their work beyond the sum of everyone's work.

Addressing food waste has been a key priority for Los Angeles since 2007. It was then when the City adopted its Solid Waste Integrated Resource Plan, a Zero Waste plan with the goal of diverting 70% of waste by 2015. In 2010, the City launched the Restaurant Food Waste Recycling Program, a waste hauler recycling rebate program. This voluntary program collects food waste as often as six days per week to be processed by a City-certified food waste processor. According to a representative from the City of Los Angeles Bureau of Sanitation, 387 food service establishments (e.g., restaurants, bakeries, and cafeterias) are participating in the program at present, with an average monthly food waste diversion per food service establishment of three tons. Since the inception of the Restaurant Food Waste Recycling Program, the City has diverted 238,000 tons of food waste from landfills (Zeuli and Nijhuis 2017). Food waste is now a key priority area for the Good Food Policy Council.

Particularly Los Angeles' approach, setup, and goals of their Good Food Policy Council can inform Greater Miami's work, as none of the three counties has a food policy council. Further, their work on reusing food and avoiding food waste is noteworthy for Greater Miami's community nutrition resilience strategy.

4.6 Other Cities

100 Resilient Cities has identified three other cities in the 100RC network that are dealing with challenges related to food security: El Paso, Bristol, and Durban. These will be discussed below.

In the USA, El Paso is working to increase the resilience of their shared and limited water supply and fragile desert ecology together with their Mexican neighbor Juarez. Together, the cities are addressing immigration, and working to increase the resilience of their shared and limited water supply (100ResilientCities.org. 2018a). As El Paso is not directly working on food system nor community nutrition resilience, it won't be reviewed in any more detail.

Outside of the USA, the City of Bristol, UK, applies an approach to deal with food security challenges that is interesting for several reasons. The city has developed an action plan as part of their resilience strategy that aims to respond to the main stresses and risks of acute shocks it is facing. One of them is food insecurity (100ResilientCities.org. 2018b). One of their actions to make the city's food system more resilient is to apply Transformative Leadership (Bristol City Council 2018).

According to the City's strategic plan, Transformative Leadership is a practical leadership framework that seems promising when dealing with food system. Aiming to train and build skills in adaptive leadership to drive collaboration and co-creation, it helps take on the gradual but meaningful process of change, both individually and collectively. As previously discussed, food systems, with elements including growing, processing, logistics, distribution, selling, buying, cooking and eating, are complex and in order to drive change, new leadership and skills will be required. This type of leadership empowers citizens and promotes long-termism, as well as protecting the natural environment and promoting health and well-being through a transformed food system.

This project, led in collaboration between the Bristol City Council, the Food Policy Council, Bristol Green Capital, MIT U-Theory, and Transition Bristol, is developing and testing approaches to delivering more resilient approaches, including U-Theory. U-Theory, or Theory U, is a change management method developed by Otto Scharmer. It is based on the theory that a shift to a higher state of attention allows individuals to operate from a future space of possibility that they feel wants to emerge. Leadership's role in this theory is to facilitate this shift to a more open will, mind, and heart (Scharmer 2019). Given the complexity of challenges related to food system resilience and community nutrition resilience, both Transformative Leadership and U-Theory seem helpful for leaders to apply when trying to bring about change in Greater Miami.

The third and final city that self-certifies as facing food insecurity within the 100RC network is Durban, South Africa (100ResilientCities.org. 2018c). Food security was identified as one of seven key themes to be addressed in a citywide climate change adaptation and mitigation strategy (Environmental Planning & Climate Protection Department 2014).

As part of the food security strategy, the City of Durban initially developed a vision and five goals. They then developed an extensive list of strategies to achieve each of these aims (Environmental Planning & Climate Protection Department 2014), including, for example, farmers' trainings in sustainable agriculture, biogas generation from food waste, and education on healthy eating habits. These vision, goals, and strategies will inform the proposed way forward for enhancing community nutrition resilience in Miami, as discussed in Chapter 5 of this book.

4.7 Food Waste and Recovery

Although not a learning from one city per se, it should be noted that food waste reduction and food recovery is gaining momentum in work on community nutrition resilience. Besides the work of the cities discussed above, Madison, Wisconsin, for example, has adopted food waste as a new priority for the Madison Food Policy Council and the Food Policy Coordinator. A seven-member food waste reduction taskforce, consisting of members from the Madison Food Policy Council and Dane County Food Policy Council, is reviewing city practices in the area of food waste to identify composting partners and stakeholders that could partner on food waste reduction efforts (Zeuli and Nijhuis 2017).

In Portland, former Mayor Michael Brennan formed the Mayor's Initiative for a Healthy and Sustainable Food System in 2012 to support initiatives to improve the health and sustainability of the city's food system, including reducing food waste. Currently known as Shaping Portland's Food System, it will soon become the Portland Food Council. The group's policy subcommittee is beginning to investigate policy interventions that could reduce food waste in Portland, including incentives for processing or recycling food waste, enabling redistribution from city institutions and initiating citywide composting. The subcommittee is chaired by an attorney from the Conservation Law Foundation, an environmental advocacy organization based in New England that is working to address food waste throughout New England. Proposed state legislation has focused on reducing food waste in landfills (Zeuli and Nijhuis 2017).

While food waste is not yet a priority for every city, interest in addressing this issue in the USA and globally suggests that city-level food waste reduction programs will continue to expand. In June 2016, the United States Conference of Mayors adopted a resolution to strengthen food waste reduction initiatives within cities. In addition, with support from The Rockefeller Foundation, the National Resources Defense Council is delving into the waste streams in New York City, Denver, and Nashville, measuring the amount of food that is wasted in homes, businesses, and large institutions and how much of this is edible and could be prevented or recovered. Measuring the amount of food that is wasted is a critical first step for municipal governments as they design strategies to divert edible food from landfills. The project will also generate a toolkit of policies and programs that other cities in the USA can use to advance their own food waste prevention, rescue and recycling efforts. There are a growing number of partnerships between municipalities and local institutions, such as universities, to use wasted food as an input into incubators for businesses working on "upcycled" products or community composting programs (Zeuli and Nijhuis 2017).

REFERENCES

100ResilientCities.org. (2018a). *El Paso's resilience challenge* [online]. Available at: http://www.100resilientcities.org/cities/el-paso/. Accessed 14 January 2019.
100ResilientCities.org. (2018b). *Bristol's resilience challenge* [online]. Available at: http://www.100resilientcities.org/cities/bristol/. Accessed 14 January 2019.
100ResilientCities.org. (2018c). *Durban's resilience challenge* [online]. Available at: http://www.100resilientcities.org/cities/durban/. Accessed 14 January 2019.
100ResilientCities.org. (2018d). *Boston's resilience challenge* [online]. Available at: http://www.100resilientcities.org/cities/boston/. Accessed 14 January 2019.
Baltimore City. (2018). *2018 food environment brief* [online]. Available at: https://planning.baltimorecity.gov/sites/default/files/City%20Map%20Brief%20011218.pdf. Accessed 15 January 2019.
Baltimorecity.gov. (2019). *Department of planning: Healthy food environment strategy* [online]. Available at: https://planning.baltimorecity.gov/baltimore-food-policy-initiative/food-environment. Accessed 15 January 2019.
Bristol City Council. (2018). *Bristol resilience strategy* [online]. Available at: http://www.100resilientcities.org/wp-content/uploads/2017/07/Bristol_Strategy_PDF.compressed.pdf. Accessed 14 January 2019.
Castro, C. (2018, November 20). Personal interview.
City of Orlando. (2019). *2018 community action plan* [online]. Available at: https://beta.orlando.gov/NewsEventsInitiatives/Initiatives/2018-Community-Action-Plan. Accessed 10 January 2019.
Environmental Planning & Climate Protection Department. (2014). *Durban climate change strategy: Food security theme report: Draft for public comment* [online]. Available at: http://www.durban.gov.za/City_Services/energy-office/Documents/DCCS%20Food%20Security%20Theme%20Report.pdf. Accessed 15 January 2019.
ICIC.org. (2015a). *Resilient food systems, resilient cities: Recommendations for the city of Boston. Executive summary* [online]. Available at: http://icic.org/wp-content/uploads/2016/04/ICIC_Food_Systems_ExecutiveSummary_final.pdf. Accessed 15 January 2019.
ICIC.org. (2015b). *Resilient food systems, resilient cities: Recommendations for the City of Boston. Implementation roadmap.* [online] Available at: http://icic.org/wp-content/uploads/2016/04/ICIC_Food_Systems_Roadmap_final_v2.pdf. Accessed 15 January 2019.

Los Angeles Food Policy Council. (2019). *Join the good food movement.* [online] Available at: https://www.goodfoodla.org/. Accessed 13 February 2019.

NYC Food Policy. (2019a). *Welcome* [online]. Available at: https://www1.nyc.gov/site/foodpolicy/index.page. Accessed 13 February 2019.

NYC Food Policy. (2019b). *About* [online]. Available at: https://www1.nyc.gov/site/foodpolicy/about/nyc-food-policy.page. Accessed 13 February 2019.

NYC Food Policy. (2019c). *Food metrics report* [online]. Available at: https://www1.nyc.gov/site/foodpolicy/about/food-metrics-report.page. Accessed 13 February 2019.

Scharmer, O. (2019). *Addressing the blind spot of our time: An executive summary of the new book by Otto Scharmer Theory U: Leading from the future as it emerges* [online]. Available at: https://www.presencing.org/assets/images/theory-u/Theory_U_Exec_Summary.pdf. Accessed 13 February 2019.

Stanford Social Innovation Review. (2011). *Collective impact* [online]. Available at: https://ssir.org/articles/entry/collective_impact. Accessed 25 January 2019.

Zeuli, K., & Nijhuis, A. (2017). *The resilience of America's urban food systems: Evidence from five cities* [ebook]. Roxbury, MA: ICIC. Available at: http://icic.org/wp-content/uploads/2017/01/Rockefeller_ResilientFoodSystems_FINAL_post.pdf?x96880. Accessed 18 December 2017.

CHAPTER 5

Conclusions—Making Greater Miami's Communities Nutrition Resilient

Abstract This final chapter provides recommendations on how to strengthen community nutrition resilience in Greater Miami. Based on interviews, literature review, and case study research, it offers actionable ideas in two areas: (1) planning and policy and (2) strengthening the food system. Several steps are identified to assign clear responsibility and focus on the issue, conduct robust assessments of critical communities on which to focus, and create the business case for multi-sectorial action. Recommendations are then made for the process of strengthening the food system, particularly increasing health literacy and cultural relevance of food for an increased uptake of healthy food options. Finally, it offers a vision for Greater Miami as a center of excellence for resilience and offers areas for future research.

Keywords Greater Miami · Community nutrition resilience planning and policy · Food policy council · Strengthen the food system · Cultural relevance

As the previous discussion has demonstrated, Greater Miami overall lacks a focus on food systems and community nutrition resilience. It appears that not one organization is working on strengthening the food system and thereby making it more resilient to withstand chronic stressors as well as sudden shocks. Further, community nutrition is not a focus of any of the strategies and programs designed to make the region more resilient.

Although work on climate adaptation has started, there is a need for more consistent action across municipalities and counties and for support of cities with less budget. Some cities such as Hialeah, for example, appear not to be doing anything at the moment.

Despite an apparent lack of support on federal and state level—which would enable critical access to technology, a more consistent sharing of data and financial resources to scale efforts—much can be done on a local level.

The previous chapter has provided insights into the work of other cities on food systems and community nutrition resilience. In this final chapter, recommendations will now be presented to strengthen Greater Miami's community nutrition resilience based on these learnings, as well as on an analysis of pertinent literature (most notably Zeuli and Nijhuis 2017) and on interviews with experts in the field.

Recommendations cover two areas: (1) Recommendations related to planning and policy—as these provide the theoretical and logistical underpinnings for the work and (2) Recommendations to strengthen the food system—covering practical actions along the food chain that ultimately ensures community nutrition resilience.

5.1 Recommendations Related to Planning and Policy

5.1.1 Form a Central Entity Responsible for Community Nutrition Resilience

Any focus on community nutrition resilience must start with creating one entity that guides strategy and oversees actions in the field. The analysis conducted by Zeuli and Nijhuis (2017) on urban resilience in the USA found that cities working on nutrition resilience have government offices (e.g., Office of Food Initiatives or Director of Food Policy) directing their food priorities. The case studies discussed in Chapter 4 all have a central unit dealing with policy, planning, and implementation.

When looking at where such initiative could be located, the City of Baltimore's approach seems best suited for various reasons. Baltimore's Food Policy Initiative is a collaboration between the Department of Planning, Office of Sustainability, Baltimore City Health Department, and Baltimore Development Corporation (Baltimorecity.gov 2019). It makes sense that these four areas should drive a community nutrition resilience focus, as all four are needed to build a sustainable, resilience, healthy, and equitable food system serving local communities.

And even if there is not one office dedicated to this cause, a food initiative, food policy council or similar must exist to promote and oversee activities in the space. Food policy councils that are independent public–private partnerships could be especially effective, since they can be somewhat detached from ongoing resilience planning and implementation and changing administrations and mayoral priorities which can disrupt the process (Zeuli and Nijhuis 2017).

In fact, the current lack of food policy councils in Greater Miami on either county or municipal level is a huge impediment to the efforts to strengthen food system resilience and community nutrition resilience. It is the recommendation of the author to initially establish three food policy councils on county level that could support municipalities. In the absence of a position of Director of Food Policy, these councils could be co-led by the County's Planning and Zoning Departments and the Offices of Sustainability and Resilience.

The food policy councils should include stakeholder representing multiple sectors and organizations. In order to effectively incorporate food systems into the cities' planning and resilience initiatives, representatives from all steps of the food system (including production, processing, packaging, distribution/access, consumption, and recovery) should be included in this effort. This will ensure that all facets of this complex system are explored and connections across the entire food system are strengthened (Zeuli and Nijhuis 2017). There could be standing members of the Greater Miami's councils that include representative organizations from along the food system, and additionally, working groups could be established in which members of the community could contribute around projects and programs in these areas. Focus would be on recommending necessary changes to current land use and comprehensive plans as well as with regard to alleviating the enforcement of these codes, felt by some foodpreneurs that were interviewed for this book as "harassment" where support from city or county level was needed.

Since the Office of Sustainability and Resilience would co-lead the council, resilience would be an integral part of the council's work. In most cities in the USA, food policy priorities typically include improving food security, managing nutrition assistance programs, promoting healthy eating and perhaps managing farmers' markets, city gardens and other urban agriculture, but not food system resilience (Zeuli and Nijhuis 2017). Including food system resilience on the agenda of Greater Miami's food policy councils would create an efficient platform to advance resilience policies and initiatives and help them avoid supporting initiatives that may unintentionally create a more vulnerable food system (Zeuli and Nijhuis 2017).

5.1.2 Conduct Assessment of the Status Quo of Community Nutrition Resilience

In order to work on the areas of highest priority and thus ultimately improve community nutrition resilience in the region, one first needs to analyze the status quo. A starting point has been attempted in Tables 2.1 and 2.2. However, considerable new research will be needed to fill existing gaps. For example, Feeding South Florida is currently working on two new studies. One aims to map food insecurity on neighborhood level in Greater Miami and the other looks at the correlation between poor nutrition and the academic performance of school children. However, a more systematic and comprehensive study covering parameters related to food security, community health and nutrition resilience on neighborhood level for the three counties is needed. This could be supported, for example, by the Department of Health and Human Services, which funded Feeding America's report "A Healthier Future for Miami-Dade County. Expanding Supermarket Access in Areas of Need" in 2017.

For this future analysis, in addition to the parameters presented in Tables 2.1 and 2.2, communities should additionally be evaluated based on the two parameters: resource robustness and adaptive capacity (Longstaff et al. 2010). Resource robustness thereby is determined by (1) resource performance, or how well the resource accomplishes a particular goal; (2) resource redundancy, or how many "back-up" resources a community possesses that can fulfill the same function; and (3) resource diversity, or how many different resources a community possesses to achieve a goal. Adaptive capacity, on the other hand, is comprised of (1) institutional memory, or the accumulated shared experience and local knowledge within the community; (2) innovative learning, the ability of the community to use information to create novel strategies of adaptation; and (3) connectedness, or the links between community members, the strength of which determines how well institutional memory and innovative learning gets diffused throughout the community. An assessment of a community's resilience can be done appropriately by examining the level of these six proposed traits related to resources and adaptive capacity of the community's critical sub-systems (Longstaff et al. 2010). For each neighborhood, it should hence be analyzed if the resources related to the food system (e.g., food production and access points, transportation routes into the neighborhood, cooling system backups) are robust and if residents are adaptive to changes in the food system, particularly if they share experiences and can reach out to one another in the case of disturbances.

Finally, it is recommended to conduct a risk perception analysis in the communities to ascertain the level of awareness about and concern related to health risks related to nutrition insecurity. As argued by Withanachchi et al. (2018), this can increase active engagement in situational analysis and active governance within the neighborhoods. This study could be done through self-certification in those communities with the greatest needs and resilience challenges.

5.1.3 Map the Food System, Identify Health Disparities, and Support the Business Case

A necessary second step to analyze the status quo of community nutrition resilience is a mapping of healthy food assets such as grocery stores that offer nutritious food, as well as complimentary healthy food access points like farmers' markets, Community-supported agricultures (CSAs), or vegetable prescription points. This effort could be made in collaboration with an academic institution such as the University of Miami and their School of Architecture or with Miami Center for Architecture & Design (AIA). Another key partner could be the Health and the Built Environment Committee of the Consortium for a Healthier Miami-Dade, which aims to educate the community, public, and private stakeholders about the health impacts of the built environment, develop strategies and influence solutions (Healthymiamidade.org 2019). As a result of above asset mapping, food deserts can be identified which should be in focus for action.

A related topic to consider is terminology, and how to generate support for the efforts. As discussed earlier, the City of Baltimore has introduced the term "Healthy Food Priority Areas" instead of "food deserts." This is because "Food desert" suggests there is no food, when in fact there is an imbalance between healthy and unhealthy food. Additionally, it does not acknowledge any grassroots work occurring to fill the gaps. Positioning community nutrition resilience efforts in a positive way seems beneficial to gather support amongst Greater Miami's communities and leaders.

After mapping the food system, including its main assets, health disparity maps should be created, covering the same areas. These two sets of maps can then be overlaid to identify the neighborhoods with the greatest need for action. Figure 5.1 is an example of such health disparity map, displaying lower-income communities with low supermarket sales and high rates of death from diet-related diseases. It is an example of the type of maps that need to be created for communities in all three counties.

Finally, the challenge of community nutrition resilience should be translated into economic terms to appeal to policy makers that allocate budgets. By applying average health-related costs to those parts of the population identified as in poor health and correlating these back to poor access to healthy, nutritious food, cumulated costs per neighborhood due to low community nutrition resilience can be derived. These form the basis for a business case for intervention programs such as the creation of community gardens, improved product offerings in food pantries or mobile farmers' markets that accept Supplemental Nutrition Assistance Program (SNAP) benefits.

To illustrate this, James Jiler (2018), Executive-Director of Urban Green Works (UGW), gives an example calculation. UGW is a Miami-based nonprofit organization which provides environmental programs and green job training to incarcerated men and women, youth remanded by court to drug rehab and at-risk high-school youth in low-income neighborhoods. Jiler summarizes the business case on the example of keeping one person out of prison by providing her with an occupation and/or attachment to her community in the form of a community garden. Costing about US$60,000 per year to incarcerate one person, this is in stark contrast to establishing a garden for US$20,000, which benefits the whole community.

5 CONCLUSIONS—MAKING GREATER MIAMI'S ...   133

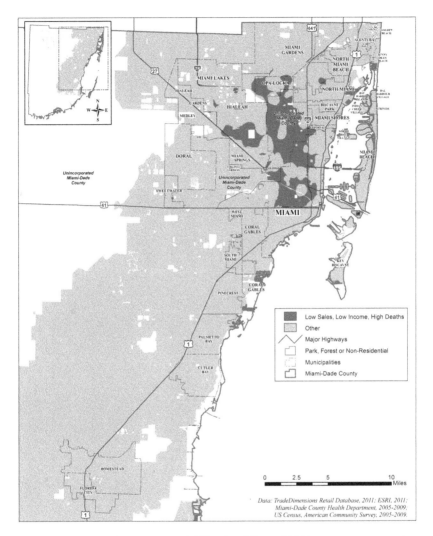

Fig. 5.1 Areas with greatest need (The Food Trust 2012)

5.1.4 Conduct Food System Resilience Assessments

After mapping the food system and defining the areas of greatest need and the focus of budget allocation based on a business case analysis, vulnerabilities that affect assets related to each step of the food system should be identified. Table 1.1 has summarized the main challenges to the resilience in food systems due to an increase in temperature and severe weather events. This overview is useful when conducting such assessment in Greater Miami.

Table 5.1 gives examples of some additional challenges to consider at each of the steps along Greater Miami's food system. This is based on the methodology offered by Zeuli and Nijhuis (2017) for looking at food system vulnerabilities. Also, Longstaff et al. (2010) proposed set of attributes to assess the resilience of community food system has been integrated in the overview.

However, a full vulnerability assessment is needed, forming the basis for the work on community nutrition resilience. Potential partners to conduct such assessment could be John's Hopkins Center for a Livable Future, Rockefeller Foundation, or Initiative for a Competitive Inner City (ICIC).

As Greater Miami is already investing in infrastructure improvements to make their cities more resilient overall to natural disasters, a food system resilience study could help Miami prioritize investments that directly

Table 5.1 Challenges to Greater Miami's food system

Food system step	Challenges
Food production	• Increased competition from Mexico • Decrease in yield due to heat stress • Increased weed pressure • Increased disease pressure • Saltwater intrusion due to sea-level rise (SLR) • Lack of talent
Food processing Food distribution	• Plant buildings vulnerable to flooding and/or storm damage • Need to be able to accept (electronic) food assistance benefits • Food retailers do not prefer local food in their sourcing • Higher costs of cooling due to more heat days and hotter climate
Food consumption	• Decrease in cooking skills and preference for food products that focus on convenience over health benefits • Different products are culturally relevant in different communities • Commodification of food and lack of connection leads to low support for local and sustainable food • Low health literacy

impact the food system. These include improving the primary roads and bridges used for food distribution, identifying alternative food supply pathways that could include use of ports and railways, and strengthening, protecting, or moving food distribution facilities out of "at risk" locations.

5.1.5 Develop a Vision, Strategy, and Aims for Community Nutrition Resilience

Based on the above-discussed assessment of the status quo, a vision and goals for community nutrition resilience in the face of chronic stresses and acute shocks should be formulated. Based on the goals, strategies to achieve them can be devised.

Below is a suggestion for Miami-Dade County (MDC) as an example, derived from looking at the work of other cities such as Los Angeles (Los Angeles Food Policy Council 2019), Durban (Environmental Planning & Climate Protection Department 2014), and Palm Beach County (2015).

Example of Miami-Dade's Community Nutrition Resilience vision, mission, and goals:

Vision
Miami-Dade's residents have a robust and resilient food security status in terms of availability, access, and nutritional benefits of food in the face of a warming climate.

Mission
Promote healthy and sustainable food in Miami-Dade County, thereby supporting MDC's second-largest industry and making sure that Miami's communities at all times have physical and economic access to sufficient, safe, and nutritious food that meets their dietary needs and food preferences for an active and healthy life. This food originates from efficient, effective, and low-cost food systems that are compatible with a sustainable use of natural resources.

Goals
1. Miami-Dade has a robust, sustainable, equitable, and efficient food system that is able to withstand future climate threats.
2. Miami-Dade has adequate food distribution and marketing networks (physical access to food) in place to provide residents with

access to healthy and nutritious food at all times, as well as to adapt to climate change.
3. Miami-Dade residents have economic access to healthy and nutritious food in the face of chronic environmental, social, economic, and climatic stresses and acute shocks.
4. Miami-Dade residents are able to utilize food in the best possible manner in the light of chronic stresses and acute shocks.
5. Miami-Dade is prepared for climate-related disasters or events and is able to supply its residents with adequate food during these disasters.

The final vision, goals, and strategies, however, should be defined in collaboration with multiple stakeholders from diverse backgrounds working along the food system in Greater Miami's counties and cities.

5.1.6 Incorporate Food Systems and Community Nutrition Resilience into Resilience Planning

Community nutrition resilience should be included in the existing resilience planning initiatives in the region. This includes the Southeast Florida Regional Compact for Climate Change. A future update should bring more focus on this area. A first step would be to include recommendations in either the Public Health or Agriculture category, or alternatively underneath Sustainable Communities and Transportation covering infrastructure-related improvements. Further, the Resilient305 strategy—Greater Miami and the Beaches' (GM&B's) plan based on their involvement with 100 Resilient Cities (100RC)—should consider food system resilience. Finally, Miami's membership in 100RC should be extended to focus on food security.

Another example of how to integrate food systems into ongoing resilience planning and thereby strengthen community nutrition resilience is through the resilience hubs. Planned by Miami-Dade County, for example, as Community Operations Centers that are potentially self-sufficient (off-grid) from an energy perspective and house community services, in future include a food element such as cooling hubs from where food can be distributed post disruption. The goal of these hubs would be to get communities to be self-sufficient for at least 3–4 weeks in case of hurricanes. It will need to be ensured, however, that nutritional quality of food is equally considered as quantity.

The Beacon Council's community ventures program pursues a similar concept and goal as the resilience hubs: to provide solutions to challenges that prolong long-term unemployment. The centers, which are currently in piloting phase, would provide wraparound services such as childcare, health care, and solutions to mobility issues to mitigate barriers of entry into the workforce. The centers are planned in partnerships with, for example, United Way, and Career Force South Florida. Finally, Catalyst is working on the concept of resilience hubs to provide social services all year around. They do that by running community visioning workshops, which supports, again, the idea of co-creation and ensures that these centers deliver what the communities need.

The main challenge related to these hubs is finding available land. To address this challenge, collaboration is needed and Opportunity Zones (OZ) could be explored. As there are many overlaps in terms of focus on particular neighborhoods, collaboration between Miami-Dade County and organizations working on the same concept is recommended, particularly as MDC owns land. OZ could offer an interesting way of securing private sector investment for these hubs. These were established as part of the Tax Cuts and Jobs Act of 2017 to spur economic development and job-creation activity in low-income communities. They work by investors receiving a variety of incentives and tax benefits for investing capital gains in low-income communities (Beacon Council 2019). The creation of resilience hubs could be one of the community investment vehicles for Greater Miami, where developers would build mixed-use, mixed-income properties that include emergency services, as well as space for local food production, health literacy education and/or healthy food access points. Thus, these hubs would address food security challenges in the face of both acute shocks and chronic stresses.

5.1.7 Co-create Neighborhood Nutrition Resilience Plans

Once the communities with the highest need are identified, prioritizing neighborhoods where nutrition resilience is particularly low, or food access would be disproportionately impacted by a natural disaster, planning can begin.

In the planning efforts, programs and initiatives should be prioritized that increase both the robustness (made up of performance, diversity, and redundancy) of local resources such as food access points, disposable income, or local leadership. Further, they should focus on improving the adaptive capacity of the local community, defined as institutional memory, the ability to learn and innovate, and the connectedness within the community.

Most importantly, community nutrition resilience plans should be co-created by listening and responding to the needs of the respective community. Thereby, it is ensured that plans address the specific local needs and engagement with the plan is fostered. Specific actions to strengthen neighborhoods by increasing the resilience of the local food system will be discussed under Sect. 5.2.

5.1.8 Develop Government Policies and Practices that Help the Food System Quickly Return to Normal Operations

After an acute shock, policies and practices should be in place to speed up the recovery of food system players. Three government policies are critical for food business recovery—food safety inspections, the construction permit process, and transportation restrictions (Zeuli and Nijhuis 2017).

Related to food safety inspections, streamlining the process is key. After a disaster, food businesses may be unable to resume operations until passing a food safety inspection. This process, coupled with limited resources for inspections, may lead to delays in the reopening of food businesses. Government agencies should develop a protocol for streamlining the food business inspection and construction permit process in the aftermath of a disaster and for effectively communicating the requirements to every food business. In addition, a process should be developed for identifying additional inspectors with the appropriate training, who can be quickly mobilized to ensure all inspections are completed in a timely manner (Zeuli and Nijhuis 2017).

When it comes to construction permits, focus should be put on speeding up permits for rebuilding efforts of damaged facilities. Finally, transportation restrictions should be eliminated where possible. An example is prohibiting food distribution trucks from entering impacted areas, slowing the distribution of food immediately after a disaster. State governments also should have a policy in place for coordinating with the federal government to temporarily suspend federal Hours of Service regulations for food distribution drivers in the aftermath of a disaster.

The regulations may be temporarily suspended under declared states of emergencies for drivers providing vital supplies and transportation services to a disaster area. State governments should also pass legislation that designates food distributors and owners of food businesses as "essential" to emergency recovery. In April 2016, Florida passed legislation (SB 1288—Post-Disaster Re-Entry) enabling businesses that provide "essentials in commerce" to transport their products during a declared emergency. Under this legislation, the Florida Division of Emergency Management will develop a certification system and permit certain activities by certified drivers or employers during a curfew, and authorize law enforcement officers to specify permissible routes for certified persons in a declared disaster area. Food retailers and distributors would be eligible for this certification. Its supporters, including the Florida Retail Federation, anticipate that the law may allow for faster distribution of food supplies in the aftermath of a disaster, potentially decreasing recovery time and thereby improving Florida's food system resilience (Zeuli and Nijhuis 2017).

Additionally, communication during and in the aftermath of a disaster is key. Efficiently and effectively sharing information between government agencies and private sector food businesses can help businesses return to normal operations as quickly as possible. Businesses may be confused about whom to contact for relevant and timely information and, in turn, government agencies may not know the best way to effectively share information. Trade associations, which typically operate at a state level, could provide a single point of contact for government agencies. Some state emergency management offices may already have mechanisms in place to coordinate with food retail trade associations during natural disasters to improve emergency response and recovery efforts. Such partnerships help with disaster recovery (e.g., providing donated food and water supplies) and identify the resources businesses need to prepare for disasters and quickly return to normal operations (e.g., identifying transportation routes or other infrastructure requiring maintenance). Finally, associations can also help to catalyze food retailers to establish business continuity plans and assist them with resilience planning. City leaders should establish similar relationships with food retailers at the local level, including them in local emergency management planning. Further, not all grocery stores, especially small grocery stores and corner stores, are members of these trade associations. Non-member stores need to have established direct lines of communication with government agencies during and in the aftermath of a natural disaster (Zeuli and Nijhuis 2017).

Another key topic is providing funding for business recovery. Government agencies can provide capital to business owners to help them rebuild and reopen for business. State and local government can also marshal resources to support the recovery of food businesses that lack sufficient capital to reopen. Often smaller grocery and corner stores that were severely damaged are slow to reopen in part because of a lack of financial resources and insufficient or delayed insurance payments. Small business recovery loan programs could alleviate this challenge (Zeuli and Nijhuis 2017). Florida has established a Small Business Emergency Loan Program to support small businesses after disasters. The program, managed by the Florida Department of Economic Opportunity provides short-term, interest-free loans to small businesses that experience physical or economic damage from a disaster to help bridge the gap between the time damage is incurred and when a business secures other financial resources, including insurance claims and long-term loans (Zeuli and Nijhuis 2017).

On a consumer level, governments are essential in ensuring food insecure households have access to D-SNAP benefits and opportunities to use Women, Infants and Children (WIC) benefits during and after a disaster. However, as federal and state regulations are less relevant to food systems, local leadership particularly from Mayors and Commissioners, who pass these regulations, is critical to take action.

5.2 Recommendations to Strengthen Food System Resilience

Below recommendations will be discussed to increase the resilience of all steps along the food system, from food production, over processing, distribution, and access (retail and consumption). Further, opportunities related to food recovery will be looked at.

Following Tyler and Moench's (2012) framework of urban resilience, looking at systems, agents, and institutions, we should be pursuing three main goals:

1. Strengthen systems to reduce their fragility in the face of acute shocks and systemic stresses and reduce the risk of cascading failure,
2. Build the capacity of social agents to anticipate and develop adaptive responses, to access and maintain a supportive system, and

3. Address the institutional factors that constrain effective responses to system fragility or undermine the ability of agents to take action.

Limited financial resources could be cited as an impediment to working on community nutrition resilience in Greater Miami. However, much work can be done inexpensively, through better policies, new ways of doing things or the promotion of behavioral change. An example is South Miami's overhaul of regular pesticide application in public places, which was replaced by on-demand spraying with less toxic substances. When financial support is needed, grants should be considered from local foundations such as the Miami Foundation or the Health Foundation of South Florida.

5.2.1 Food Production

Although only a small percentage of produce consumed in Greater Miami has been produced in the region, this section will look at how to strengthen local food production to increase food system resilience by decreasing reliance on external food sources, in addition to lowering the environmental footprint of consumed food.

5.2.1.1 Support Local Food Farming

Local food farming can alleviate many of the resilience challenges that food production is facing. As discussed earlier, agriculture in Greater Miami faces challenges such as consolidation, shrinking profitability, and a lack of future talent. These challenges are exacerbated by the consequences of climate change. Supporting local agriculture supports the second largest industry in MDC as dollars are spent locally with people that spend locally again. And although it is recognized that Miami will continue to be dependent on national and global food systems for some products and months of the year, an increased focus on strengthening local agriculture promotes the self-sufficiency of the region.

To strengthen farmers' resilience, Porter et al. (2014) offer some coping responses. These can include altering cultivation and sowing times, crop cultivars and species, and marketing arrangements. Early sowing is facilitated using techniques such as dry sowing, seedling transplanting, and seed priming. Also, later sowing and fast-growing crop varieties can help reduce exposure to end-of-season droughts and high-temperature events. Further, approaches are needed that integrate climate forecasts such as digital farming. Warmer conditions will also require improving cultivar tolerance to high temperature, improving gene conservation and excess to extensive gene banks. Changing pest, disease, and weed threats as well as possible droughts will require new crop protection and seed technology (Porter et al. 2014).

Further, investment in increased innovation will be needed to ensure the adaptation of local agriculture and the associated socioeconomic system can keep pace with climate change. Support can be lent by the local Institute of Food and Agricultural Sciences (IFAS) and other research institutions. In addition, innovations in growing methods should be investigated, such as related to farming occurring not only on land but also in the cities, and more specifically within or upon buildings. Jenkins et al. (2015), for example, looked at soil-less growing technologies using nutrient-rich water instead: hydroponics and aquaponics. Hydroponic systems thereby use nutrients added to recirculated water systems, whereby aquaponics utilize waste ammonia produced by fish as fertilizer crops in a symbiotic system (Jenkins et al. 2015). Both systems allow growing in a protective environment with crops sheltered from storms, temperature soars, or other shock events. As this system requires considerable amounts of freshwater, it needs to be investigated if the benefits outweigh the costs especially with freshwater becoming more scarce. Further, the available surface area within and upon buildings in Greater Miami would need to be determined to estimate harvest potential. However, for certain crops such as lettuce or cabbage, soil-less farming technologies can offer promising complimentary and resilient food production systems (Jenkins et al. 2015).

As mentioned above, water is another issue that commands greater attention in the area of agriculture. Agriculture, food security, and nutrition are all highly sensitive to changes in rainfall associated with climate change (Porter et al. 2014). Overall, the availability of fresh, high-quality water has important implications for the food safety, and thus quality, of agricultural produce, as well as for public health. To address these challenges, the International Water Resources Association (IWRA) has compiled a Global Compendium on Water Quality Guidelines, examining examples of existing recommendations for influent water quality management, as applied to various human and ecosystem uses (IWRA 2019). This compendium can provide a useful starting point for policy makers in Greater Miami aiming to safeguard or improve water quality.

In Miami, particularly the restoration of the Everglades and of Biscayne Bay is key to providing freshwater in the future. At the same time, both provide important natural buffers from flooding and protect critical ecosystems (Resilient305.org 2018). Further investments into infrastructure to stop saltwater intrusion into wells are needed to protect freshwater supplies, hence, securing access to irrigation water for farming, safeguarding its quality for domestic and industrial use, and protecting critical ecosystems in and around it.

In addition, the local agriculture industry could benefit from a stronger organization, such as, for example, in the local Farm Bureau. This would help with advocacy for local needs, including the need for labeling or other marketing efforts, with cost-sharing, promoting ag innovation and generally having a voice as, for example, on the North American Free Trade Agreement (NAFTA) trade agreement that impacts farmers' profitability.

Equally important is to strengthen the agricultural industry with fresh talent. Due to its vast ethnic mix of immigrants, many families in Greater Miami originally stem from countries with vast agricultural communities, and there is an opportunity to bring back their children into the industry. Opportunities here are multi-language communication campaigns that portray agriculture as interesting and trendy, work experience programs for high schoolers on farms, or incentives to join agricultural training programs. Another way of increasing economic resilience of farmers and to prevent that they give up farming entirely is through an increased diversification outside of agriculture such as with farm tourism, or the production of processed food items.

Most importantly, in order to strengthen local farming, it needs to be promoted with consumers. According to Sage (2014), the localization and revitalization of new food production and consumption networks help to recover natural endowments of each region, help connect locals to ecology and seasonality, and generally become the most feasible, practicable, and frequently successful first step toward food security. In fact, it can increase food security on national level as previously discussed. As a diverse local agriculture ensures continued redundancies, it would directly increase food system resilience in Greater Miami. In addition, it makes people eat more fresh produce, which increases overall health.

There are several ways in which local agriculture can be promoted. First, work should be done on educating people to make a conscious choice. This can include increased labeling such as the "Fresh from

Florida" or "Redland Made." Similarly, efforts should be made to work with retailers to offer more local food, thereby decreasing transportation costs, increasing freshness and offering peak taste. The promotion of Community Supported Agriculture (CSA) is another way of supporting local agriculture. Here, farmers would benefit from stronger bridges to local food businesses and institutions such as restaurants, or schools. Further, farmers markets are a great tool since consumers hear about where their food comes from, and consequently connect more to it instead of seeing it as a commodity.

Finally, there should be more stringent checks from policy makers or other sources on businesses claiming to buy local. For instance, there may be restaurants that make that claim but don't regularly order from local farmers. An example of how this could be publicized is the multipart investigative reporting "Farm to Fable" by the Tampa Bay Times (2019), which exposes restaurants and farmers' markets that claim to sell local produce but in reality do not.

5.2.1.2 Urban Agriculture

Urban agriculture can alleviate threats to nutrition resilience and help communities increase their connection to food. Growing food in the community opens a space to challenge the mainstream food system by offering a more equitable, ecologically sustainable and potentially socially empowering alternative (Sage 2014). It can thereby contribute to the necessary process of decommodification of nourishment addressed earlier (Wilson 2012).

Ways of incorporating agriculture into the urban landscape are via community gardens, vertical gardens, green roofs. In Miami, it can be also incorporated in the recently launched Underline project, which will transform the land below Miami's Metrorail into a 10-mile linear park, urban trail, and living art destination (Theunderline.org 2019).

Relatively easy and cost-effective to implement, community gardens can potentially offer high returns on investment when looking at health care costs in underprivileged neighborhoods, or at their role in keeping people out of prison, as discussed in Sect. 5.1.3. Community gardens create jobs, urban greening, and dietary improvements (Sage 2014). As one interviewee has argued, they hence should be considered as resources for the community, just like a library or a community service center and thus should be funded and run by government. However, as Tecco et al. (2017) have found in their analysis of community gardens in Turin, Italy, urban garden management is far from being unproblematic. Often being regarded as symbols for, or catalysts of urban change and social activism and inclusion, they need to exist at the intersection of conservative regulations and societal pressures asking for social change. Hence, their effectiveness to provide a solution to address community nutrition resilience might be challenged.

In the context of urban agriculture, edible landscapes are another interesting concept. The turfgrass found in lawns, parks, and schoolyards represents the single largest irrigated crop in the USA. Across the country, turf uses up 34 billion liters (nine billion gallons) of water per day in irrigation, demanding 31 million kilograms (70 million pounds) of pesticides and 757 million liters (200 million gallons) of gasoline annually. With an emphasis on native perennials and food-producing plants, edible landscapes can be a great way to create green space and provide healthy, fresh food. Replacing just a fraction of traditional lawn with edible landscapes designed around locally appropriate plants would have numerous benefits. Edible landscapes often require little or no additional irrigation or fertilizer can increase food production potential in cities and can be a boon to pollinators and ecological diversity (Foodtank 2019).

Community gardens also need to fit the fabric of the local community. Not all neighborhoods might appreciate an opportunity to farm. Some might find it too much work; some might even find it offensive. Hence, cultural relevance needs to be considered related to food production as much as for its consumption.

However, the main challenge related to urban agriculture in Greater Miami is that in most cities there are no supported urban agriculture policies to support zoning for small farms to grow in the urban core. Changing this, at least by issuing temporary permits until the benefits for the respective neighborhood are demonstrated, can eliminate this barrier.

Further, in the interviews for this book the challenge of scalability of urban agriculture was addressed a few times. Finally, local agriculture is not in itself a response to the issue of food availability after a hurricane as this can destroy urban landscapes. It thus merits further research.

5.2.2 Food Processing and Distribution

There is very little to none food processing done in Greater Miami locally. However, for those facilities that are located here, food packing, storage, and distribution policies and systems may need to be enhanced

(Porter et al. 2014). Further climate adaptation of buildings such as improved drainage or storm shutters as well as investment into insurance and early warning systems can alleviate threats related to severe weather.

5.2.3 Food Access

Related to food access, work can be done to increase the resilience of traditional access points such as grocery stores. It also covers work with institutions such as schools or food banks. Moreover, complimentary access points such as farmers' markets or CSA offer opportunities to strengthen resilience on food access level. Finally, one must look at challenges related to the consumption of healthy, nutritious food. All of these are discussed below.

Related to **traditional food access points**, the lack of supermarket access and increased incidence of diet-related diseases in lower-income neighborhoods suggest the need for incentive programs and policies to support healthy food retail development in underserved areas. Figure 2.3 illustrates the areas in Miami-Dade County with lower-income and low supermarket sales due to few or no supermarket locations. In Miami-Dade, these lower-income areas with a lack of food access points are concentrated in Opa-Locka, unincorporated Miami-Dade County, inner city Miami, and parts of Hialeah, Homestead, and Florida City (The Food Trust 2012).

Investments into healthy food retail development would have a positive impact on both community health and economic development as food retail markets bring jobs to neighborhoods that need them the most (The Food Trust 2012).

The availability of food access points does not allow conclusions about the healthfulness of the offered products. However, a mapping of food access in Greater Miami is a necessary first step and would still need to be completed for the counties Broward and Palm Beach. Hence, a further examination of the availability of healthful and fresh produce in these neighborhoods is critical, as the emphasis of policy and practices should be on the provision of food that addresses health and well-being related challenges in these communities.

In sum, work on food access on neighborhood level should include a long—as well as short-term view. In the long-term, two key underlying causes of neighborhood food vulnerability: Food insecurity (and the poverty that produces it) and a lack of food retail stores, especially grocery stores, need to be addressed (Zeuli and Nijhuis 2017).

In the short-term, making sure grocery and corner stores have adequate business continuity plans and insurance should be priorities (Zeuli and Nijhuis 2017). In fact, supply chain resilience is a food industry priority and many larger food businesses already have business continuity plans in place. Smaller food companies, however, may be underprepared for business disruptions and may have inadequate business continuity plans and insurance coverage in place. These businesses may not be clear about the potential impact a disaster could have on their business, they may not believe that a disaster is likely to happen, or they may not fully understand their insurance policies. Greater Miami's cities should work with the food industry to review business continuity plans and insurance coverage for all food businesses to gain insight into their plans and help them to address any shortcomings. Trade associations can also help to catalyze and provide resources to their members to improve business continuity plans. For example, Food Marketing Institute, a national trade association for the food retail industry, provides resources and holds forums for sharing best practices on business continuity, crisis management, and building organizational resilience (Zeuli and Nijhuis 2017). Further, incubators such as StartUp Florida International University (FIU) Food, which focuses on helping individuals start and scale local food businesses, should prepare these small businesses for the chronic stressors and acute shocks present in Greater Miami and discussed earlier.

Looking at New York City offers another potential route. The city has gone a step further and called on their state legislature to mandate larger food retailers (20,000 square feet or more of floor space or 60 or more full- or part-time employees) to install electric generators or to make sure that they have the ability to hook up to a mobile generator or other alternative power source to ensure that food retailers have power to process transactions and operate emergency lighting and fire and security systems during a disaster. As of 2016, this legislation had not yet passed (Zeuli and Nijhuis 2017). Although Florida lawmakers as of 2018 require nursing homes and assisted living facilities to have generators, there is yet no law that requires food retailers to have any.

Engaging the food chain is a necessary part of the solution to enhancing community nutrition resilience. Programs can include, for example, corner-store initiatives where large food retailers donate display shelves to corner stores. Further, teaching corner stores that carry fresh produce that proper shelving, cleanness, and tidiness are important for people to buy produce will be beneficial.

With regard to **schools**, the analysis of the status quo in Greater Miami has shown that uptake of breakfast programs and summer meal programs is low compared to the uptake of the school lunch program. According to the Health Foundation of South Florida, the national breakfast program, currently has low participation rates in MDC and is the biggest opportunity especially in food deserts. Several organizations such as Florida Impact, a state-level advocacy organization to increase participation in federal programs works on school programs, after school, summer break spots, or the previously introduced FLIPANY are working on these areas.

Food banks, dubbed the backbone of urban food safety nets (Zeuli and Nijhuis 2017), play a big part in community nutrition resilience. They are likely to experience a sustained increased in demand from a greater number of food insecure households well after the disaster, which would create a significant challenge for many food banks that struggle to meet the needs of existing food insecure populations. Hence, all food banks in the Feeding America network have adopted a disaster preparedness plan.

A common criticism related to food banks and pantries is that they may not carry enough healthy food items. Since their business model is to consolidate donations from the United States Department of Agriculture (USDA), farms, and corporations they often accept donations of unhealthy items that they then need to hand out. In addition, they advise customers to follow a diet based on the USDA MyPlate recommendations, the health benefits of which has been challenged as discussed in Sect. 2.4.3. Finding solutions how to secure healthier food donations, for example, through gleaning, disposing unhealthy food items in a sustainable manner (e.g., through composting), as well as employing nutrition experts to advise customers would help address these concerns. State and local governments should work with food banks and the private sector to develop a plan for securing more donations of healthy food and establishing funding for food banks to support more food purchases.

Supporting **complimentary food access points** such as farmers markets and CSA can be done by many means. On policy level, for example, by making the permitting process easier and less costly. According to

Heiman (2018), particularly impact fees are prohibitively expensive. Further, enabling test permitting as an easy and fast alternative to a permanent permit, would help in setting up innovative food concepts.

Further, working with farmers that sell via farm stands and help them enable the SNAP doubling program (FAB) for fresh produce on farmers' markets would make a difference to lower-income communities. Also related to SNAP, in the aftermath of a disaster, retailers may not have all the products eligible for WIC benefits and are since they are not allowed to substitute items, this makes it difficult for retailers to hand out food. Reviewing this restriction in the aftermath of a disaster would alleviate this bottleneck for food insecure families. In addition, SNAP and D-SNAP benefits are issued through an Electronic Benefits Transfer (EBT) card, meaning telecommunication networks need to be functioning in order to process benefits. Reconnecting food access points that accept EBT first can help re-establishing food supply for food insecure households.

Finally, on the **consumption** side, the local purchasing behavior is an important barrier to offering healthy produce in corner stores. Unlike in other cities such as NYC or Philadelphia, Miami residents don't buy fresh produce in corner stores (Schoepp 2018). Partially that is because corner stores don't highlight or sell enough fresh produce. Other challenges related to consumption behavior and possible solutions are discussed in the next section.

5.2.4 *Health Literacy and Community Education*

Literacy and communication are at the core of engaging communities in taking action on strengthening nutrition resilience in Greater Miami. Several interview partners have mentioned that it appears to be difficult to create demand for fruits and vegetables. Even if new access points like mobile farmers markets are being introduced into former food deserts, people don't want the fresh produce.

First, it is often difficult for people to connect the dots between poor health and the food they consume. So, working on their health literacy is key. Overall, coordinated efforts should be made with multiple partners working on access and most importantly including elements of health education, like for example, the Live Healthy Miami Gardens program. Moreover, it is necessary to start teaching health education at an early age at school. Further, doctors should be talking about the link between nutrition and lifestyle-related diseases.

Another important barrier to people choosing healthy nutritious food is that it is associated with not tasting as good. In Miami Gardens, for example, this was identified by one of the most important issues by the community. Community cooking demonstrations can thereby help introduce tasty recipes with healthful food items to community members.

A third barrier is the lack of knowledge what to do with the fresh produce as many people lack knowledge about fresh produce items as well as cooking skills. School gardens thereby can play a big role in educating children in food items and teach cooking skills. Further, adult cooking classes or any other activity that promotes bonding over food can be included in community education programs. Supportive environments and communities are fundamental in shaping people's choices, by making the choice of healthier foods and regular physical activity the easiest choice (the choice that is the most accessible, available, and affordable), and therefore preventing overweight and obesity (WHO.int 2018).

5.2.5 *Cultural Relevance*

A related issue to literacy is cultural relevance. As initiatives in other cities have demonstrated, if corner stores are stocked with foods people in the community don't like, produce will not be bought. This is particularly important to realize for Miami's eclectic immigrant communities. Zelalem Adefris, Resilience Director at Catalyst Miami notes, for example, that stores should rather stock coconuts, hot peppers, or mangos instead of kale and apples (Adefris 2018). Knowing which produce is liked by the community requires outreach on the part of stores, which should have an interest in selling their produce.

5.2.6 *Food Waste and Recovery*

Diverting edible food from landfills can increase the availability of meals for the food insecure. Further, recovering non-edible food and disposing of it in a sustainable manner (e.g., through composting) can create environmental benefits. Both would strengthen Greater Miami's resilience.

A possible starting point would be to establish food waste reduction goals in the municipalities, like Los Angeles has done. Their program aims to reduce landfill disposal by 1 million tons per year by 2025 and reduce waste by 65% in all 11 of the City's new service zones. It also aims to decrease food waste and provide all Angelenos with Blue Bin access, no matter where they live or work (Lacity.org 2017). These goals should be complemented by support that bridges the gap between institutions and organizations that donate and those that need the donations (e.g., food banks).

However, the net impact of food recovery initiatives on food donations remains to be seen. Food waste reduction goals could increase donations to food banks, but this could be offset by increased supply chain efficiencies that lead to less surplus food. However, food banks should be partnering with schools, food processors, and retailers to increase the efficiency of the food donation process, which means that more surplus food should make its way to food banks and pantries. Further, like Broward County, the other counties via their food policy councils could be working with the school districts, as well as the private sector to recover food and direct it to those in need.

5.3 The Way Forward: From Fun in the Sun to Resilience Excellence

According to Otis Rolley, Managing Director North America of 100RC, Miami has the chance to set an example for the world on how to deal with sea level rise. Several other interview partners have mentioned that MDC is pursuing the opportunity, or vision, to turn the county into a synonym for resilience.

Right now, the region is charting a course to combat SLR, and the efforts are on the world stage. The Chief Resilience Officer (CRO) of Miami Beach is interviewed almost on a weekly basis from people all over the world on the topic, who want to tour the city to learn from local experience (Torriente 2018). Miami Beach's response to SLR and coastal erosion could become a benchmark for global cities vulnerable to these very issues (Urban Land 2018).

In this sense, Greater Miami could become a center of excellence for adaptation technology as several other geographies with the same topography could be dealing with the same climate-related issues in future. According to a representative of the Beacon Council (Jaap 2018), this requires the fostering of innovation, e.g., driven by the local universities, and entrepreneurship. Once the expertise and solutions are acquired, they can provide an important revenue stream for local ventures that could assist other places in the world.

However, Greater Miami's current resilience efforts are incomplete as they only have recently started to consider social issues such as gentrification and are still lacking focus on food systems and community nutrition. A challenge is that the region is seen primarily as a tourism destination (sun and fun), and this industry remains important to the local tax base. Climate adaptation protects this industry in which 12% of the local workforce are employed. Further, a comprehensive focus on resilience, one that includes nutrition resilience in the face of climate change, could lead to high-paying jobs, and valuable research, tying together the private sector and academia.

However, to manage this transition in mind-set and setup of current programs, hence moving Greater Miami toward community nutrition resilience requires leadership and that this has been lacking was mentioned several times during the interviews for this book. It has been the almost unanimous opinion that Greater Miami needs bolder visionaries and leaders that look at it long-term, not in terms of political terms, and that walk the talk themselves. As an example of a local leader in climate change mitigation, the house of Mayor Philip Stoddard in South Miami is powered by solar power, as he aims to show that the technology has arrived to go house micro-grid. The Mayors in Greater Miami, on city- and county-level overall need to be stronger engaged.

What is found in Miami at the moment is great planning, but as Seville (2009: 11) puts it "blue-print planning is no substitute for great leadership and a culture of teamwork and trust which can respond effectively to the unexpected." And related to this culture of teamwork, leaders in Greater Miami need to do a better job in bringing more people from diverse sectors into the discussion room. What has started with the Regional Compact and continued in various workshops during the 100RC affiliation needs to continue and be expanded to strengthen collaboration between government entities and organizations, as well as to involve community groups and the private sector.

According to Mayor Stoddard (2018), the change that is needed has to be made possible through a back and forth between elected officials and community activists. Politicians in that case need to stronger engage the public in decision-making, for example, through committees or councils. This however requires confidence and a willingness to be transparent, even if not everything has been figured out yet.

Further, the change that is needed will be exponential, not incremental. Two methods that can help develop the type of leadership needed in Miami are Collective Impact and Transformative Leadership, both of which have been discussed in Sects. 4.6 and 4.7, respectively.

Overall, urban planning in Greater Miami needs to be more cohesive if the future footprint of the cities were to solve their main resilience challenges. The current setup of Miami, for example, requires people to live in one neighborhood, work in another (sometimes beyond county lines), and consume food that is produced somewhere else. A future center of resilience excellence would center around the idea of living hubs, thereby improving the environmental footprint, overall resilience and the well-being of people living in the region.

5.4 Areas for Further Research

Areas for further research are plenty as this book has only been able to scratch the surface of community nutrition resilience, a relatively new concept altogether, in Greater Miami. Some of them are discussed below.

First, not in scope of this discussion has been **recovery after an emergency**. This should be further looked at from a standpoint of community nutrition. Areas to discuss would include how to best make nutritious food available to communities after a shock, particularly looking at the role of food banks, how urban ag could alleviate any shortages, and how food recovery can reduce shortages in neighborhoods post-disaster.

Further, it would need to look at **how to pay for community resilience** efforts, including, e.g., who pays for diversity and redundancy in the food system (Wilkinson 2012) in order to have food available from multiple sources, how to help food businesses increase their resilience, how to provide for cultural relevance as well as community education and trainings.

Equity and justice are another area that has come up increasingly in the research for this book. Shamar Bibbins, Senior Program Officer for Environment at The Kresge Foundation, in her speech at the 2018 Annual Southeast Florida Regional Climate Leadership Summit, noted that few actors today understand how to incorporate equity and justice into adaptation and the same is true for conceptualizing how community nutrition resilience impacts equity and vice versa.

Given the global nature of our modern food system, research is also needed to understand the **impact of a natural disaster in food-exporting cities on food supply to other cities**. For example, much of the vegetables consumed in Greater Miami come from California. More research is needed to understand the type of impact interruptions in remote locations has on the supply of food to the Greater Miami market.

Equally, more research is needed to fully understand the interplay between **food waste reduction** and food system and community nutrition resilience, as well as the feasibility and potential impact of urban agriculture in Greater Miami. Finally, efforts should be made to **quantify the impact of grassroot programs** on community nutrition resilience.

5.5 Conclusions

Resilience is about more than infrastructure; it is about people. People are the fabric of cities and communities. Keeping communities strong is just as much a climate imperative as maintaining our built systems and protecting our natural systems.

Communities are resilient to chronic stresses and acute shocks when they are endowed with robust resources that are performant, redundant, and diverse and have the capacity to adapt. Hence, preparing communities does not only include making resources physically available, but also engaging and empowering residents to make use of them through education and a seat on the table.

Nutrition has a big impact on the physical, mental, emotional, and even economic development of individuals and communities. And although the challenge of providing all residents of Greater Miami with sustainable, healthy and nutritious food at all times seems large and strategies to address it complex, doing so is critical to make communities thrive and people be able to reach their full potential.

Climate change disproportionately affects vulnerable populations and discrepancies are exacerbated in a warming climate. In fact, Greater Miami has community members already facing urgent issues. As this book has shown, Greater Miami is at the moment particularly focused on resilience to environmental challenges, adapting to them mainly through infrastructure adaptation. It is less so looking at the social and economic perspectives of resilience, community nutrition resilience being one of them.

And because Greater Miami is so young, it appears that mandates are still lacking and processes are still in its infancy. It still addresses resilience through blue-print planning, instead of through collaborative and innovative approaches. Further, much action is driven by grassroot organizations and community activism.

There's no doubt that the challenges that Greater Miami is facing are enormous and that funding is limited. Moreover, better policy at State and Federal level would facilitate action. However, there is a lot more leaders in the region could be doing. Current limitations thus not only relate to policy and finance, but also to overcoming inertia. And this does not only apply to local government. Big institutions also need to understand their role outside of their traditional roles to first of all tackle resilience for the future, and secondly to make the link between sub-systems like the food system and resilience. Diversity of actors, programs, and actions plays a major role in resilience on both natural and socio-ecological levels against the effects of climate change on food and nutrition security. Related to the challenge of funding, programs that address community nutrition resilience should be inherently economically sustainable, as opposed to philanthropic.

This book can at best provide a snapshot of what Greater Miami is doing on community nutrition resilience. Programs are appearing and evolving with every strategy update, new elected official and often through advocacy of outstanding community leaders. However, it remains clear that a focus on nutrition resilience is a necessary starting point for the immediate future. Climate change will only increase the occurrence and severity of more extreme weather, rising sea levels and natural disasters in cities in the USA (Zeuli and Nijhuis 2017). And these will continue to exacerbate existing disparities in Greater Miami. Further, as this book highlights, natural disasters could create extended food supply disruptions, especially in neighborhoods with limited food retail options and food insecure populations. Hence, city leaders need to start thinking about not only their short-term response to extreme events, but also how to strengthen critical food systems long-term.

And though not discussed in detail, only time will tell if and when Greater Miami will reach a tipping point where action needs to shift from focusing on adaptation to transformation. Current strategies however should already consider difficult topics such as planned retreat. It is the hope of the author that this initial research on urban food system resilience will catalyze leaders in the region to begin to address food system vulnerabilities in their resilience planning and programs and start regarding community nutrition resilience as an essential part of ensuring the well-being and success of people living in Greater Miami for the near future.

References

Baltimorecity.gov. (2019a). *Department of planning: Healthy food environment strategy* [online]. Available at: https://planning.baltimorecity.gov/baltimore-food-policy-initiative/food-environment. Accessed 15 January 2019.

Beacon Council. (2019). *New opportunity zones could be a new beginning for Miami's inner-city* [online]. Available at: https://www.beaconcouncil.com/new-opportunity-zones-could-be-a-new-beginning-for-miamis-inner-city/. Accessed 1 February 2019.

Environmental Planning & Climate Protection Department. (2014). *Durban climate change strategy: Food security theme report: Draft for public comment* [online]. Available at: http://www.durban.gov.za/City_Services/energy-office/Documents/DCCS%20Food%20Security%20Theme%20Report.pdf. Accessed 15 January 2019.

Foodtank. (2019). *15 organizations creating edible landscapes* [online]. Available at: https://foodtank.com/news/2018/07/organizations-creating-edible-landscapes/. Accessed 25 January 2019.

Healthymiamidade.org. (2019). *Health and the built environment* [online]. Available at: https://www.healthymiamidade.org/committees/health-and-the-built-environment/. Accessed 2 March 2019.

IWRA. (2019). *New IWRA report on "Developing a global compendium on water quality guidelines"!* [online]. Available at: https://www.iwra.org/waterqualityreport/. Accessed 10 May 2019.

Jenkins, A., Keeffe, G., & Hall, N. (2015). Planning urban food production into today's cities. *Future of Foods: Journal on Food, Agriculture and Society, 3*, 35–47.

Lacity.org. (2017). *City council passes zero waste LA program* [online]. Available at: https://www.lacity.org/blog/city-council-passes-zero-waste-la-program. Accessed 2 March 2019.

Longstaff, P. H., Armstrong, N. J., Perrin, K., Parker, W. M., & Hidek, M. A. (2010). Building resilient communities: A preliminary framework for assessment. *Homeland Security Affairs, VI*(3), 1–23.

Los Angeles Food Policy Council. (2019). *Join the good food movement* [online]. Available at: https://www.goodfoodla.org/. Accessed 13 February 2019.

Palm Beach County. (2015, October). *Hunger relief plan Palm Beach County* [online]. Available at: http://discover.pbcgov.org/communityservices/humanservices/PDF/News/Palm_Beach_FRAC_100515_Edition-v2.pdf. Accessed 25 January 2019.

Porter, J. R., et al. (2014). Food security and food production systems. In *Climate change 2014: Impacts, adaptation, and vulnerability. Part A: Global and sectoral aspects. Contribution of Working Group II to the Fifth Assessment Report of the Intergovernmental Panel on Climate Change* (pp. 485–533).

Resilient305.org. (2018). *Resilient Greater Miami & the beaches: Preliminary resilience assessment #Resilient305* [online]. Available at: http://resilient305.com/assets/pdf/170905_GM&B%20PRA_v01-2.pdf. Accessed 29 January 2018.

Sage, C. (2014). The transition movement and food sovereignty: From local resilience to global engagement in food system transformation. *Journal of Consumer Culture, 14*(2), 254–275. https://doi.org/10.1177/1469540514526281.

Seville, E. (2009). Resilience: Great concept…but what does it mean for organisations? In Ministry of Civil Defence & Emergency Management (Ed.), *Community resilience: Research, planning and civil defence emergency management*. Wellington, NZ: Ministry of Civil Defence & Emergency Management.

Tampa Bay Times. (2019). *Farm to fable* [online]. Available at: http://www.tampabay.com/projects/2016/food/farm-to-fable/. Accessed 13 February 2019.

Tecco, N., Coppola, F., Sottile, F., & Peano, C. (2017). Urban gardens and institutional fences. *Future of Food: Journal on Food, Agriculture and Society, 5*(1), 70–78.

The Food Trust. (2012). *A healthier future for Miami-Dade County: Expanding supermarket access in areas of need* [online]. Available at: http://thefoodtrust.org/uploads/media_items/miami-dade-supermarket-report.original.pdf. Accessed 26 January 2019.

Theunderline.org. (2019). *Home* [online]. Available at: https://www.theunderline.org/. Accessed 13 February 2019.

Torriente, S. (2018, October 24). Personal interview.

Tyler, S., & Moench, M. (2012). A framework for urban climate resilience. *Climate and Development, 4*(4), 311–326. https://doi.org/10.1080/17565529.2012.745389.

Urbanland.uli.org. (2018). *Living with rising sea levels: Miami Beach's plans for resilience* [online]. Available at: https://urbanland.uli.org/sustainability/living-rising-sea-levels-miami-beachs-plans-resilience/. Accessed 22 October 2018.

WHO.int. (2018). *Obesity and overweight: Key facts* [online]. Available at: https://www.who.int/news-room/fact-sheets/detail/obesity-and-overweight. Accessed 16 December 2018.

Wilkinson, C. (2012). Urban resilience: What does it mean in planning practice? In S. Davoudi, K. Shaw, L. J. Haider, A. E. Quinlan, G. D. Peterson, C. Wilkinson, H. Fünfgeld, D. McEvoy, L. Porter, & L. Porter (Eds.), Resilience: A bridging concept or a dead end? "Reframing" resilience: Challenges for planning theory and practice interacting traps: Resilience assessment of a pasture management system in Northern Afghanistan urban resilience: What does it mean in planning practice? Resilience as a useful concept for climate change adaptation? The politics of resilience for planning: A cautionary note. *Planning Theory & Practice, 13*(2), 324–328. https://doi.org/10.1080/14649357.2012.677124.

Wilson, A. D. (2012). Beyond alternative: Exploring the potential for autonomous food spaces. *Antipode, 45*(3), 719–737. https://doi.org/10.1111/j.1467-8330.2012.01020.x.

Withanachchi, S., Kunchulia, I., Ghambashidze, G., Al Sidawi, R., Urushadze, T., & Ploeger, A. (2018). Farmers' perception of water quality and risks in the Mashavera River Basin, Georgia: Analyzing the vulnerability of the social-ecological system through community perceptions. *Sustainability, 10*(9), 3062.

Zeuli, K., & Nijhuis, A. (2017). *The resilience of America's urban food systems: Evidence from five cities* [ebook]. Roxbury, MA: ICIC. Available at: http://icic.org/wp-content/uploads/2017/01/Rockefeller_ResilientFoodSystems_FINAL_post.pdf?x96880. Accessed 18 December 2017.

Appendix A: List of Interview Partners (Alphabetic Order)

Zelalem Adefris, Catalyst Miami, Resilience Director, October-25, 2018.
Yoca Arditi-Rocha, The CLEO Institute, Executive Director, October-1, 2018.
Irela Bagué, Bagué Group, Principal, November-14, 2018.
José Caceres, SAP, Head of Communications Operations, Latin America and the Caribbean, October-30, 2018.
Juan P. Casimiro, BIZNOVATOR, Founder, October-22, 2018.
Chris Castro, Sustainability & Resilience at City of Orlando, Director, November-20, 2018.
Daniella Levine Cava, Miami-Dade County, District 8, Commissioner, November-19, 2018.
Jaap Donath, Research & Strategic Planning at The Miami-Dade Beacon Council, Senior Vice President, November-8, 2018.
Mary-Stewart Droege, City of Orlando, Project Manager Downtown Development Board/Community Redevelopment Agency, January-9, 2019.
Art Friedrich, Urban Oasis Project, President, October-22, 2018.
Jane Gilbert, City of Miami, Chief Resilience Officer, October-10, 2018.
Jill Horwitz, Broward County Environmental Planning and Community Resilience Division, Natural Resource Specialist, December-11, 2018.
Megan Houston, Office of Resilience at Palm Beach County, Director, October-17, 2018
James Jiler, Urban GreenWorks, Co-founder & Executive-Director, November-15, 2018.
Elisa Tatiana Juarez, Target, Corporate Responsibility, November-1, 2018.

© The Editor(s) (if applicable) and The Author(s), under exclusive license to Springer Nature Switzerland AG 2020
F. Alesso-Bendisch, *Community Nutrition Resilience in Greater Miami*, Palgrave Studies in Climate Resilient Societies, https://doi.org/10.1007/978-3-030-27451-1

Jennifer Jurado, Broward County, Chief Resiliency Officer & Director, October-17, 2018.
Charles LaPradd, Miami-Dade County, Agricultural Manager, November-2, 2018.
Taruna Malhotra, Palm Beach County, Deputy Director Community Services, January-9, 2019.
James Murley, Miami-Dade County, Chief Resilience Officer, November-26, 2018.
Jeannie Necessary, UF IFAS Extension Family Nutrition Program, Public Health Specialist, October-22, 2018.
Anthony Olivieri, FHEED, Founder, June-14, 2018.
Janisse Rosario-Schoepp, Health Foundation of South Florida, Vice President of Operations and Strategy, June-4, 2018.
Solina Rulfs, UF IFAS Extension Family Nutrition Program, Food Systems, January-11, 2019.
Gretchen Schmidt, Edible South Florida, Writer, November-7, 2018.
Thi Squire, Grow2Heal, Homestead Hospital, Baptist Hospital, Project Manager, November-19, 2018.
Kate Stein, WLRN, Climate and environment reporter, December-4, 2018.
Philip Stoddard, City of South Miami, Mayor, September-21, 2018.
Mary-Stewart Droege, City of Orlando, Downtown Development Board/ Community Redevelopment Agency, Project Manager, January-9, 2019.
Susanne M. Torriente, City of Miami Beach, Chief Resiliency Officer & Assistant City Manager, October-24, 2018.
Paco Vélez, Feeding South Florida, President and CEO, January-22, 2019.

Appendix B: RCAP Actions by Municipality (Southeastfloridaclimatecompact.org 2018d)

Appendix B: RCAP Actions by Municipality ...

Category name	Abbreviated title	Recommendation text	Miami-Dade					Broward					Palm Beach			
			Miami Beach	City of Miami	Hialeah	Miami Gardens	South Miami	Fort Lauderdale	Pembroke Pines	Hollywood	Miramar	Coral Springs	Pompano Beach	West Palm Beach	Boca Raton
Agriculture	AG-1	Promote policies that preserve the economic viability of agriculture									x				
	AG-2	Continue to meet the water needs of agriculture											x		
	AG-3	Promote locally produced foods and goods					x	x			x			x	
	AG-4	Align research and extension with climate-related needs of agriculture									x				
	AG-5	Maintain or create agriculture purchase of development rights programs													
	AG-6	Assess opportunities for growers and agricultural land-owners to manage land to lessen the impacts of climate change and incentivize those actions				x									

Category name	Abbreviated title	Recommendation text	Miami-Dade					Broward					Palm Beach			
			Miami Beach	City of Miami	Hialeah	Miami Gardens	South Miami	Fort Lauderdale	Pembroke Pines	Hollywood	Miramar	Coral Springs	Pompano Beach	West Palm Beach	Boca Raton
	AG-7	Seek a national designation for Southeast Florida as a critical source of domestic agricultural products													
	AG-8	Identify and reduce obstacles for enabling urban agriculture, gardening, and other backyard agricultural practices						x			x				
	AG-9	Increase resources for the study and implementation of invasive, non-native pest and pathogen prevention; early detection; and rapid response													
	AG-10	Promote sustainable aquaculture, perennial crops, diversified farming systems, precision agriculture, and re-contouring field elevations													

			Miami-Dade				Broward					Palm Beach			
Category name	Abbreviated title	Recommendation text	Miami Beach	City of Miami	Hialeah	Miami Gardens	South Miami	Fort Lauderdale	Pembroke Pines	Hollywood	Miramar	Coral Springs	Pompano Beach	West Palm Beach	Boca Raton
	AG-11	Assess and address public health risks of more frequent and intense high-heat days to agriculture and farm workers													
Compact coordination	CC-1	Establish and implement a regional communications strategy amongst business, government, and community leadership													
	CC-2	Update regional unified sea level rise projections													
	CC-3	Explore opportunities to better coordinate cross-agency and cross-jurisdiction reviews of major infrastructure projects													

Category name	Abbreviated title	Recommendation text	Miami-Dade					Broward						Palm Beach			
			Miami Beach	City of Miami	Hialeah	Miami Gardens	South Miami	Fort Lauderdale	Pembroke Pines	Hollywood	Miramar			Coral Springs	Pompano Beach	West Palm Beach	Boca Raton
	CC-4	Continue to provide high-quality implementation support and resources for jurisdictions seeking to implement RCAP recommendations and other sustainability and resilience measures															
	CC-5	Develop and track regional indicators of climate change impacts, emission reduction, and adaptation action															
	CC-6	Create a Compact advisory group composed of organizations that represent the region's climate work, equitable community development, and vulnerable populations in order to track and share best practices on equitable climate action with the region															

			Miami-Dade				Broward					Palm Beach			
Category name	Abbreviated title	Recommendation text	Miami Beach	City of Miami	Hialeah	Miami Gardens	South Miami	Fort Lauderdale	Pembroke Pines	Hollywood	Miramar	Coral Springs	Pompano Beach	West Palm Beach	Boca Raton
Energy and fuel	EF-1	Promote renewable energy through policies and technological development in order to reduce greenhouse gas (GHG) emissions	x			x	x	x	x	x	x		x		
	EF-11	Establish a fuel-efficient municipal vehicle fleet													
	EF-12	Promote community use of electric vehicles (EV)													
	EF-2	Advance energy efficiency and conservation through technological solutions, behavioral strategies, and policies in order to reduce greenhouse gas (GHG) emissions		x				x	x	x	x		x	x	
	EF-6	Streamline permitting and administrative processes to reduce the soft costs associated with renewable energy technologies													

APPENDIX B: RCAP ACTIONS BY MUNICIPALITY ...   167

Category name	Abbreviated title	Recommendation text	Miami-Dade					Broward				Palm Beach			
			Miami Beach	City of Miami	Hialeah	Miami Gardens	South Miami	Fort Lauderdale	Pembroke Pines	Hollywood	Miramar	Coral Springs	Pompano Beach	West Palm Beach	Boca Raton
	EF-7	Establish financing mechanisms for current homeowners to invest in renewable energy and energy efficiency				x									
	EF-9	Enable grid-independent energy and waste-to-energy systems			x				x	x	x		x		
	EF-10	Enable a fuel-efficient public vehicle fleet						x	x	x				x	
Natural systems	NS-1	Foster public awareness of the impacts of climate change on the region's natural systems and ecosystem services	x					x		x					
	NS-2	Promote collaborative federal, state, and local government conservation land acquisition and easement programs					x								
	NS-3	Support regional wildland fire management coordination efforts											x		

Category name	Abbreviated title	Recommendation text	Miami-Dade				Broward					Palm Beach			
			Miami Beach	City of Miami	Hialeah	Miami Gardens	South Miami	Fort Lauderdale	Pembroke Pines	Hollywood	Miramar	Coral Springs	Pompano Beach	West Palm Beach	Boca Raton
	NS-4	Develop sustainable financing for the monitoring, protection, restoration, and management of natural areas and ecosystem services					x								
	NS-5	Identify or create a regional group to coordinate a plan to create adaptation corridors, living collections, and other approaches to species dispersal and conservation						x						x	
	NS-6	Conduct a predictive assessment of current and potential invasive species ranges and impacts		x		x	x		x	x			x	x	
	NS-7	Promote the protection and restoration of coastal natural systems and the creation of living shorelines at the regional scale	x											x	

APPENDIX B: RCAP ACTIONS BY MUNICIPALITY ... 169

			Miami-Dade					Broward				Palm Beach			
Category name	Abbreviated title	Recommendation text	Miami Beach	City of Miami	Hialeah	Miami Gardens	South Miami	Fort Lauderdale	Pembroke Pines	Hollywood	Miramar	Coral Springs	Pompano Beach	West Palm Beach	Boca Raton
	NS-8	Support coral reef protection, restoration, and sustainable-use initiatives to help Florida's sensitive reefs adapt to the changing climate and ocean acidification													
	NS-9	Advocate for federal and state funding for applied monitoring and climate-related science, conducted in partnership with the Florida Climate Institute					x								
	NS-10	Examine and propose revisions to environmental regulations to account for the effects of climate change													
	NS-11	Identify the effects of climate change on fish populations, the sustainability of key fisheries, and the fishing industry, then develop adaptation plans as needed													

			Miami-Dade					Broward				Palm Beach			
Category name	Abbreviated title	Recommendation text	Miami Beach	City of Miami	Hialeah	Miami Gardens	South Miami	Fort Lauderdale	Pembroke Pines	Hollywood	Miramar	Coral Springs	Pompano Beach	West Palm Beach	Boca Raton
	NS-12	Promote the protection, restoration, and creation of freshwater wetlands, open space buffer areas, and connectivity between freshwater and estuarine waters	x	x											
	NS-13	Develop and implement long-term, sustainable, regional solutions to beach erosion and sediment supply				x									
	NS-14	Maintain, create, and/or restore urban tree canopy	x				x	x	x	x	x		x	x	
	NS-15	Support and advocate for continued implementation and funding on the state and federal levels for the Comprehensive Everglades Restoration Plan (CERP)													

APPENDIX B: RCAP ACTIONS BY MUNICIPALITY ... 171

			Miami-Dade					Broward					Palm Beach			
Category name	Abbreviated title	Recommendation text	Miami Beach	City of Miami	Hialeah	Miami Gardens	South Miami	Fort Lauderdale	Pembroke Pines	Hollywood	Miramar	Coral Springs	Pompano Beach	West Palm Beach	Boca Raton
Public health	PH-1	Understand and communicate public health risks associated with climate change													
	PH-2	Adopt and update all Florida Department of Health plans to reflect climate and sea level rise impacts on public health													
	PH-3	Adapt federal and state public health resources to support specific community needs													
	PH-4	Reduce extreme heat exposure to promote public health													
	PH-5	Advocate for policy changes and funding for local health departments to collect data more frequently to influence public health plans													

Category name	Abbreviated title	Recommendation text	Miami-Dade					Broward				Palm Beach			
			Miami Beach	City of Miami	Hialeah	Miami Gardens	South Miami	Fort Lauderdale	Pembroke Pines	Hollywood	Miramar	Coral Springs	Pompano Beach	West Palm Beach	Boca Raton
	PH-6	Increase reporting of health data monitoring systems to evaluate emerging diseases related to climate change													
	PH-7	Develop tools to assess the impacts of climate change and sea level rise on existing chronic conditions and to report trends or concerns for action													
Public outreach and engagement	PO-1	Assess community needs to guide local government communications													
	PO-2	Promote public awareness and understanding of climate impacts, as well as the personal actions and public policy options available to respond to climate change		x			x		x	x	x		x	x	
	PO-3	Inspire community action to address the causes and impacts of climate change											x	x	

			Miami-Dade					Broward				Palm Beach			
Category name	Abbreviated title	Recommendation text	Miami Beach	City of Miami	Hialeah	Miami Gardens	South Miami	Fort Lauderdale	Pembroke Pines	Hollywood	Miramar	Coral Springs	Pompano Beach	West Palm Beach	Boca Raton
	PO-4	Create regional open data platforms and digital tools													
	PO-5	Create culturally- and linguistically-appropriate information gathering tools and strategies to help inform decision-makers of the priorities and concerns in communities													
Public policy advocacy	PP-1	Support at all levels of government policy, legislation, and funding to reduce greenhouse gas emissions in all sectors, use less energy and water, deploy renewable energy and low-carbon transportation, prepare for and adapt to climate impacts, build community resilience, and study climate and earth science	x	x			x		x	x	x		x	x	

		Miami-Dade					Broward						Palm Beach		
Category name	Abbreviated title	Recommendation text	Miami Beach	City of Miami	Hialeah	Miami Gardens	South Miami	Fort Lauderdale	Pembroke Pines	Hollywood	Miramar	Coral Springs	Pompano Beach	West Palm Beach	Boca Raton
	PP-2	Develop common positions on climate, energy, and resilience issues, and advocate jointly as the Compact for those positions before state and federal legislatures, regulatory bodies, and the executive and judicial branches of government				x									
	PP-3	Urge federal, state, regional, and local partners to prioritize climate change considerations in the planning, construction, and operation of the regional water management and flood control system					x	x	x	x					
	PP-4	Participate in coalitions of public-, private-, nonprofit-, and/or academic-sector actors dedicated to climate, energy, and resilience issues		x									x	x	

APPENDIX B: RCAP ACTIONS BY MUNICIPALITY ...   175

Category name	Abbreviated title	Recommendation text	Miami-Dade					Broward						Palm Beach			
			Miami Beach	City of Miami	Hialeah	Miami Gardens	South Miami	Fort Lauderdale	Pembroke Pines	Hollywood	Miramar		Coral Springs	Pompano Beach	West Palm Beach	Boca Raton
	PP-5	Coordinate climate, energy, and resilience policies amongst counties, municipalities, school districts, and other units of government in the region														
	PP-6	Prioritize climate policies that advance social and economic equity for high-vulnerability populations and limited-income residents														
	PP-7	Consider the direct and indirect impacts of projects, policies, and investments on relevant stakeholders														
Regional economic resilience	ER-2	Advance regional resilience infrastructure standards														
	ER-3	Seek federal and state engagement to develop a resilience strategy														

			Miami-Dade				Broward					Palm Beach			
Category name	Abbreviated title	Recommendation text	Miami Beach	City of Miami	Hialeah	Miami Gardens	South Miami	Fort Lauderdale	Pembroke Pines	Hollywood	Miramar	Coral Springs	Pompano Beach	West Palm Beach	Boca Raton
Risk reduction and emergency management	RR-1	Identify and quantify infrastructure and populations at risk to sea level rise and storm surge						x		x			x		
	RR-3	Integrate climate vulnerability analysis data, as well as climate adaptation planning and funding, into existing emergency planning and funding documents		x			x	x					x		
	RR-9	Review the Florida Building Code through the lens of climate vulnerability	x					x	x				x	x	
	RR-11	Promote and leverage existing policies and programs designed to reduce flood risks and economic losses						x							
	RR-13	Use effective social media for emergency messaging, public health updates, and tidal flooding updates	x	x											

			Miami-Dade					Broward					Palm Beach			
Category name	Abbreviated title	Recommendation text	Miami Beach	City of Miami	Hialeah	Miami Gardens	South Miami	Fort Lauderdale	Pembroke Pines	Hollywood	Miramar	Coral Springs	Pompano Beach	West Palm Beach	Boca Raton
	RR-15	Support disaster planning and preparedness training for city and county staff													
	RR-17	Ensure the emergency management definition of communities at risk "includes economically vulnerable people"						x					x		
	RR-18	Align and integrate emergency management staff and responsibilities with chief resilience officer roles to bolster long-term plans													
	RR-8	Continue to adopt and update consistent plans at all levels of government in the region that address and integrate mitigation, sea level rise, and climate change adaptation				x		x	x	x	x		x		

Category name	Abbreviated title	Recommendation text	Miami-Dade				Broward					Palm Beach			
			Miami Beach	City of Miami	Hialeah	Miami Gardens	South Miami	Fort Lauderdale	Pembroke Pines	Hollywood	Miramar	Coral Springs	Pompano Beach	West Palm Beach	Boca Raton
	RR-6	Prioritize adaptation investments to reduce the impact of flooding and sea level rise on transportation infrastructure, particularly on evacuation routes				x	x	x							
	EQ-1	Encourage dialogue between elected officials, staff, and socially vulnerable populations about local climate impacts and community priorities to inform leaders of community needs								x			x		
	EQ-2	Integrate social vulnerability data into all local government processes													
	EQ-4	Address the needs of socially vulnerable populations by engaging existing community leaders and representative organizations in decision-making processes, particularly for critical public infrastructure													

Category name	Abbreviated title	Recommendation text	Miami-Dade					Broward				Palm Beach			
			Miami Beach	City of Miami	Hialeah	Miami Gardens	South Miami	Fort Lauderdale	Pembroke Pines	Hollywood	Miramar	Coral Springs	Pompano Beach	West Palm Beach	Boca Raton
	EQ-6	Partner with intermediary organizations that have deep community ties with socially vulnerable populations to co-create engagement and outreach strategies													
	EQ-7	Provide equity and social justice training for local government staff													
Sustainable communities and transportation	ST-1	Incorporate unified sea level rise projections, by reference, into all city, county, and regional agency comprehensive plans, transportation plans, and other infrastructure plans, and capital improvement plans	x	x		x	x	x	x	x	x		x	x	
	ST-2	Ensure locally produced maps for planning and project documents include the latest storm surge and sea level rise projections						x					x		

Category name	Abbreviated title	Recommendation text	Miami-Dade				Broward					Palm Beach			
			Miami Beach	City of Miami	Hialeah	Miami Gardens	South Miami	Fort Lauderdale	Pembroke Pines	Hollywood	Miramar	Coral Springs	Pompano Beach	West Palm Beach	Boca Raton
	ST-10	Employ transit-oriented developments and other planning approaches to promote higher-density development capable of supporting more robust transit		x		x	x	x		x	X		x		
	ST-11	Modify local land use plans and ordinances to support compact development patterns, creating more walkable and affordable communities				x	x	x	x	x	X		x	x	
	ST-12	Develop and implement policies and design standards that recognize the transportation system most vulnerable users and incorporate sustainable elements						x		x	x		x		

		Miami-Dade				Broward					Palm Beach			
Category name	Abbreviated title	Recommendation text	Miami Beach	City of Miami	Hialeah	Miami Gardens	South Miami	Fort Lauderdale	Pembroke Pines	Hollywood	Miramar	Coral Springs	Pompano Beach	West Palm Beach	Boca Raton
	ST-13	Conduct an assessment of unused or underutilized properties and develop an approach for utilizing such properties that enhances overall resilience goals				x	x	x	x	x	x		x	x	
	ST-16	Phase out septic systems where necessary to protect public health and water quality													
	ST-17	Ensure investments reduce greenhouse gas (GHG) emissions and increase the resilience of the transportation system to extreme weather and climate impacts	x			x	x	x	x		x		x	x	
	ST-18	Increase the use of transit as a transportation mode for the movement of people in the region		x		x		x	x	x				x	

			Miami-Dade				Broward					Palm Beach			
Category name	Abbreviated title	Recommendation text	Miami Beach	City of Miami	Hialeah	Miami Gardens	South Miami	Fort Lauderdale	Pembroke Pines	Hollywood	Miramar	Coral Springs	Pompano Beach	West Palm Beach	Boca Raton
	ST-19	Expand, connect, and complete networks of bicycle and pedestrian facilities, including those supporting access to transit			x		x	x	x		x				
	ST-20	Expand the use of transportation demand management strategies to reduce peak period and single-occupant vehicle travel						x					x	x	
	ST-22	Implement transportation system management and operations strategies to maximize the efficiency of the existing transportation system in a coordinated manner across local governments and agencies in the region						x	x					x	
	ST-3	Use vulnerability and risk assessment analyses and tools to identify priorities for resilience investments													

APPENDIX B: RCAP ACTIONS BY MUNICIPALITY … 183

Category name	Abbreviated title	Recommendation text	Miami-Dade					Broward					Palm Beach			
			Miami Beach	City of Miami	Hialeah	Miami Gardens	South Miami	Fort Lauderdale	Pembroke Pines	Hollywood	Miramar	Coral Springs	Pompano Beach	West Palm Beach	Boca Raton
	ST-4	Designate adaptation action areas, restoration areas, and growth areas as a priority-setting tool for vulnerable areas, and as a means to maximize benefits to natural systems while guiding people and commerce to less vulnerable places in the region	x			x	x	x		x	x		x	x	
	ST-6	Develop localized adaptation strategies for areas of greatest climate-related vulnerability in collaboration with appropriate agencies and jurisdictions to foster multi-jurisdictional solutions and maximize co-benefits		x		x	x	x	x	x	x	x			

Category name	Abbreviated title	Recommendation text	Miami-Dade					Broward						Palm Beach			
			Miami Beach	City of Miami	Hialeah	Miami Gardens	South Miami	Fort Lauderdale	Pembroke Pines	Hollywood	Miramar		Coral Springs	Pompano Beach	West Palm Beach	Boca Raton
	ST-7	Incorporate strategies to reduce risk and economic losses associated with sea level rise and flooding into local comprehensive plans, post-disaster redevelopment plans, building codes, and land development regulations	x			x		x		x	x				x	
	ST-8	Consider the adoption of green building standards to guide decision-making and development and to provide an incentive for better location, design, and construction of residential, commercial, and mixed-use developments and redevelopment		x						x						

APPENDIX B: RCAP ACTIONS BY MUNICIPALITY … 185

Category name	Abbreviated title	Recommendation text	Miami-Dade					Broward					Palm Beach			
			Miami Beach	City of Miami	Hialeah	Miami Gardens	South Miami	Fort Lauderdale	Pembroke Pines	Hollywood	Miramar	Coral Springs	Pompano Beach	West Palm Beach	Boca Raton
Water	WS-10	Integrate combined surface and groundwater impacts into the evaluation of at-risk infrastructure and the prioritization of adaptation improvements	x	x		x	x	x		x			x	x	x
	WS-11	Encourage green infrastructure and alternative strategies													
	WS-12	Integrate hydrologic and hydraulic models		x				x		x			x	x	
	WS-13	Practice integrated water management and planning							x	x	x				
	WS-15	Foster scientific research for improved water resource management		x				x	x	x			x	x	
	WS-16	Expand partnerships and resources to further innovation in water resource management					x	x	x	x	x		x	x	
	WS-17	Advance capital projects to achieve resilience in water infrastructure				x	x	x		x			x	x	

Category name	Abbreviated title	Recommendation text	Miami-Dade				Broward					Palm Beach			
			Miami Beach	City of Miami	Hialeah	Miami Gardens	South Miami	Fort Lauderdale	Pembroke Pines	Hollywood	Miramar	Coral Springs	Pompano Beach	West Palm Beach	Boca Raton
	WS-18	Coordinate innovation and regional funding													
	WS-19	Recognize adaptable infrastructure													
	WS-20	Support the Comprehensive Everglades Restoration Plan (CERP)					x		x		x		x		
	WS-21	Expand regional surface water storage								x			x		
	WS-1	Foster innovation, development, and exchange of ideas for managing water					x	x		x	x		x	x	
	WS-3	Plan for future water supply conditions	x			x	x			x			x	x	
	WS-4	Coordinate saltwater intrusion mapping across Southeast Florida						x		x			x		
	WS-5	Maintain regional inventories of water and wastewater infrastructure											x	x	

APPENDIX B: RCAP ACTIONS BY MUNICIPALITY ... 187

Category name	Abbreviated title	Recommendation text	Miami-Dade				Broward					Palm Beach			
			Miami Beach	City of Miami	Hialeah	Miami Gardens	South Miami	Fort Lauderdale	Pembroke Pines	Hollywood	Miramar	Coral Springs	Pompano Beach	West Palm Beach	Boca Raton
	WS-7	Modernize infrastructure development standards in the region													
	WS-8	Address the resilience of the regional flood control system													
	WS-9	Update the regional stormwater rule													

BIBLIOGRAPHY

100ResilientCities.org. (2018a). *Frequently asked questions (FAQ) about 100 resilient cities* [online]. Available at: http://www.100resilientcities.org/100RC-FAQ/#/-_/. Accessed 29 January 2018.
100ResilientCities.org. (2018b). *El Paso's resilience challenge* [online]. Available at: http://www.100resilientcities.org/cities/el-paso/. Accessed 14 January 2019.
100ResilientCities.org. (2018c). *Bristol's resilience challenge* [online]. Available at: http://www.100resilientcities.org/cities/bristol/. Accessed 14 January 2019.
100ResilientCities.org. (2018d). *Durban's resilience challenge* [online]. Available at: http://www.100resilientcities.org/cities/durban/. Accessed 14 January 2019.
100ResilientCities.org. (2018e). *Boston's resilience challenge* [online]. Available at: http://www.100resilientcities.org/cities/boston/. Accessed 14 January 2019.
ABC Local 10 News. (2014). *Report: 2/3 of Miami-Dade County residents overweight, obese* [online]. Available at: https://www.local10.com/news/florida/miami-dade/report-2_3-of-miami-dade-county-residents-overweight-obese. Accessed 21 December 2018.
Adger, W. N. (2003). Building resilience to promote sustainability. *IHDP Update* (pp. 1–3).
Aggarwal, A., Rehm, C. D., Monsivais, P., & Drewnowski, A. (2017). *Importance of taste, nutrition, cost and convenience in relation to diet quality: Evidence of nutrition resilience among US adults using National Health and Nutrition Examination Survey (NHANES) 2007–2010*. Prev. Med. Author manuscript; available in PMC 2017 September 01. 90: 184–192. https://doi.org/10.1016/j.ypmed.2016.06.030.

American Immigration Council. (2018). *Immigrants in Florida* [online]. Available at: https://www.americanimmigrationcouncil.org/research/immigrants-florida. Accessed 10 December 2018.

Armitage, D., Berkes, F., & Doubleday, N. (2007). Introduction: Moving beyond co-management. In D. Armitage, F. Berkes, & N. Doubleday (Eds.), *Adaptive co-management: Collaboration, learning and multi-level governance* (pp. 1–15). Vancouver and Toronto: UBC Press.

Baltimore City. (2018). *2018 food environment brief* [online]. Available at: https://planning.baltimorecity.gov/sites/default/files/City%20Map%20Brief%20011218.pdf. Accessed 15 January 2019.

Baltimorecity.gov. (2019a). *Department of planning: Healthy food environment strategy* [online]. Available at: https://planning.baltimorecity.gov/baltimore-food-policy-initiative/food-environment. Accessed 15 January 2019.

Baltimorecity.gov. (2019b). *Department of planning: Healthy food environment strategy* [online]. Available at: https://planning.baltimorecity.gov/baltimore-food-policy-initiative/healthy-food-retail. Accessed 15 January 2019.

Bazerghi, C., McKay, F. H., & Dunn, M. (2016). The role of food banks in addressing food insecurity: A systematic review. *Journal of Community Health, 41*(4), 732–740. https://doi.org/10.1007/s10900-015-0147-5.

Beacon Council. (2019a). *One community one goal: Preparing Miami-Dade County for long-term economic growth* [online]. Available at: https://www.beaconcouncil.com/ocog/. Accessed 20 February 2019.

Beacon Council. (2019b). *New opportunity zones could be a new beginning for Miami's inner-city* [online]. Available at: https://www.beaconcouncil.com/new-opportunity-zones-could-be-a-new-beginning-for-miamis-inner-city/. Accessed 1 February 2019.

Bec, A., McLennan, C. L., & Moyle, B. D. (2016). Community resilience to long-term tourism decline and rejuvenation: A literature review and conceptual model. *Current Issues in Tourism, 19*(5), 431–457.

Beeheavenfarm.com. (2018). *CSA: Info* [online]. Available at: http://beeheavenfarm.com/csa/info/. Accessed 19 October 2018.

Berkes, F., & Folke, C. (1998). *Linking social and ecological systems: Management practices and social mechanisms for building resilience*. Cambridge: Cambridge University Press.

Bloomberg. (2016). *Benchmark: The 10 most unequal cities in America: The South Florida city is neck and neck with Atlanta and New Orleans* [online]. Available at: https://www.bloomberg.com/news/articles/2016-10-05/miami-is-the-newly-crowned-most-unequal-city-in-the-u-s. Accessed 13 December 2018.

Bloomberg. (2017). *South Florida's real estate reckoning could be closer than you think* [online]. Available at: https://www.bloomberg.com/news/features/2017-12-29/south-florida-s-real-estate-reckoning-could-be-closer-than-you-think. Accessed 29 January 2018.

Bristol City Council. (2018). *Bristol resilience strategy* [online]. Available at: http://www.100resilientcities.org/wp-content/uploads/2017/07/Bristol_Strategy_PDF.compressed.pdf. Accessed 14 January 2019.

Broward.org. (2018). *Climate change* [online]. Available at: http://www.broward.org/Climate/Pages/default.aspx Accessed 15 October 2018.

Broward.org. (2019). *Broward County land use plan* [online]. Available at: http://www.broward.org/BrowardNext/Pages/broward-county-land-use-plan.aspx. Accessed 14 January 2019.

Brundtland, G. (1987). *Report of the world commission on environment and development: Our common future* [online]. Available at: http://www.un-documents.net/our-common-future.pdf. Accessed 6 September 2018.

Business Insider. (2018). *Miami is racing against time to keep up with sea-level rise* [online]. Available at: https://www.businessinsider.com/miami-floods-sea-level-rise-solutions-2018-4. Accessed 10 October 2018.

C40.org. (2016). *C40's executive director Mark Watts: Mayors are changing the way we think about food* [online]. Available at: https://www.c40.org/blog_posts/c40-s-executive-director-mark-watts-mayors-are-changing-the-way-we-think-about-food. Accessed 20 September 2018.

C40.org. (2017). *Urban planning and development initiative: Food systems* [online]. Available at: http://www.c40.org/networks/food_systems. Accessed 22 January 2018.

C40.org. (2018). *Special projects: Inclusive climate action* [online]. Available at: https://www.c40.org/programmes/inclusive-climate-action. Accessed 20 September 2018.

Carpenter, S. R., Westley, F., & Turner, G. (2005). Surrogates for resilience of social-ecological systems. *Ecosystems, 8*(8), 941–944. https://doi.org/10.1007/s10021-005-0170-y.

Castro, C. (2018, November 20). Personal interview.

Cdc.gov. (2017). *Food desert* [online]. Available at: https://www.cdc.gov/healthcommunication/toolstemplates/entertainmented/tips/FoodDesert.html. Accessed 16 December 2018.

Cdc.gov. (2019a). *Zika virus: Pregnancy* [online]. Available at: https://www.cdc.gov/zika/pregnancy/index.html. Accessed 13 February 2019.

Cdc.gov. (2019b) *What is health literacy* [online]. Available at: https://www.cdc.gov/healthliteracy/learn/index.html. Accessed 2 February 2019.

Census.gov. (2019a). *Quick facts Florida* [online]. Available at: https://www.census.gov/quickfacts/fl. Accessed 2 February 2019.

Census.gov. (2019b). *American housing survey—Table creator* [online]. Available at: https://www.census.gov/programs-surveys/ahs/data/interactive/ahstablecreator.html. Accessed 4 February 2019.

Center for Immigration Studies. (1995). *Shaping Florida: The effects of immigration, 1970–2020* [online]. Available at: https://cis.org/Report/

Shaping-Florida-Effects-Immigration-19702020. Accessed 10 December 2018.
Ces.fau.edu. (2014). *Risk, resilience and sustainability: A case study of Fort Lauderdale* [online]. Available at: http://www.ces.fau.edu/publications/pdfs/fort-lauderdale-case-studyv3.pdf. Accessed 16 October 2018.
Citizens' Climate Lobby. (2018). *What is the Climate Solutions Caucus?* [online]. Available at: https://citizensclimatelobby.org/climate-solutions-caucus/. Accessed 23 December 2018.
Cityage. (2018). *Susanne Torriente, Chief Resilience Officer, City of Miami Beach* [online]. Available at: http://cityage.org/project/susanne-torriente-chief-resilience-officer-city-of-miami-beach/. Accessed 22 October 2018.
City of Orlando. (2019). *2018 community action plan* [online]. Available at: https://beta.orlando.gov/NewsEventsInitiatives/Initiatives/2018-Community-Action-Plan. Accessed 10 January 2019.
CNBC Markets. (2018). *Rising risks: 'Climate gentrification' is changing Miami real estate values—For better and worse* [online]. Available at: https://www.cnbc.com/2018/08/29/climate-gentrification-is-changing-miami-real-estate-values.html. Accessed 14 September 2018.
CNN.com. (2017). *US ending 'wet foot, dry foot' policy for Cubans* [online]. Available at: https://www.cnn.com/2017/01/12/politics/us-to-end-wet-foot-dry-foot-policy-for-cubans/index.html. Accessed 2 February 2019.
Coast.noaa.gov. (2019). *Fast facts: Economics and demographics* [online]. Available at: https://coast.noaa.gov/states/fast-facts/economics-and-demographics.html. Accessed 1 March 2019.
Commonthreads.org. (2019). *Home* [online]. Available at: http://www.commonthreads.org/. Accessed 12 February 2019.
Dade County Farm Bureau. (2018). *Dade agriculture* [online]. Available at: http://www.dade-agriculture.org/p/dade-agriculture.html. Accessed 8 October 2018.
DataUSA.io. (2019). *Find a profile* [online]. Available at: https://datausa.io/search/?kind=geo. Accessed 10 January 2019.
Davoudi, S. (2012). Resilience: A bridging concept or a dead end? In S. Davoudi & L. Porter (Eds.), Applying the resilience perspective to planning: Critical thoughts from theory and practice. *Planning Theory & Practice, 13*(2), 299–307. https://doi.org/10.1080/14649357.2012.677124.
Drawdown. (2019). *Food: Reduced food waste* [online]. Available at: https://www.drawdown.org/solutions/food/reduced-food-waste. Accessed 12 February 2019.
Environmental Planning & Climate Protection Department. (2014). *Durban climate change strategy: Food security theme report: Draft for public comment* [online]. Available at: http://www.durban.gov.za/City_Services/energy-office/Documents/DCCS%20Food%20Security%20Theme%20Report.pdf. Accessed 15 January 2019.

Ericksen, P. J., Ingram, J. S. I., & Liverman, D. M. (2009). Food security and global environmental change: Emerging challenges. *Environmental Science & Policy, 12*, 373–377. https://doi.org/10.1016/j.envsci.2009.04.007.

FAO. (2008). *Climate change adaptation and mitigation in the food and agriculture sector*. Technical Background Document from the Expert Consultation held on 5 to 7 March 2008, FAO, Rome.

FAO.org. (2006). *Food security: Policy brief* [online]. Available at: http://www.fao.org/forestry/13128-0e6f36f27e0091055bec28ebe830f46b3.pdf. Accessed 22 January 2018.

FAO.org. (2018). *The state of food security and nutrition in the world* [online]. Available at: http://www.fao.org/state-of-food-security-nutrition/en/. Accessed 17 December 2018.

FAO.org. (2019a). *Key facts on food loss and waste you should know!* [online]. Available at: http://www.fao.org/save-food/resources/keyfindings/en/. Accessed 2 February 2019.

FAO.org. (2019b). *Food wastage footprint & climate change* [online]. Available at: http://www.fao.org/3/a-bb144e.pdf. Accessed 1 March 2019.

Farm Flavor. (2018). *Florida agriculture* [online]. Available at: https://www.farmflavor.com/florida-agriculture/. Accessed 8 October 2018.

Farmshare.org. (2019). *About us* [online]. Available at: http://farmshare.org/about-us/. Accessed 2 March 2019.

Feeding South Florida. (2017). *Map the meal gap 2017: A landmark analysis of food insecurity in the United States* [online]. Available at: http://dev.feedingsouthflorida.org/wp-content/uploads/2017/05/Map-the-Meal-Gap-Feeding-South-Florida-2017.pdf. Accessed 5 February 2019.

Feeding South Florida. (2018). *South Florida continues to face hunger challenges* [online]. Available at: https://feedingsouthflorida.org/south-florida-continues-to-face-hunger-challenges/. Accessed 20 February 2019.

Finkelstein, E. A., Trogdon, J. G., Cohen, J. W., & Dietz, W. (2009). Annual medical spending attributable to obesity: Payer-and service-specific estimates. *Health Affairs (Millwood), 28*(5), 822–831. https://doi.org/10.1377/hlthaff.28.5.w822.

Flipany.org. (2019). *What we do* [online]. Available at: http://flipany.org/about-flipany/what-we-do. Accessed 10 February 2019.

Florida Phoenix. (2019). *How much is climate change costing us? A bill in the Legislature would start an official accounting* [online]. Available at: https://www.floridaphoenix.com/2019/03/29/how-much-is-climate-change-costing-us-a-bill-in-the-legislature-would-start-an-official-accounting/. Accessed 13 May 2019.

FMI.org. (2019). *Food industry glossary* [online]. Available at: https://www.fmi.org/our-research/food-industry-glossary. Accessed 13 February 2019.

Folke, C. (2006). Resilience: The emergence of a perspective for social–ecological systems analyses. *Global Environmental Change, 16*(3), 253–267.
Folke, C., Carpenter, S., Walker, B., Scheffer, M., Chapin, T., & Rockstrom, J. (2010). Resilience thinking: Integrating resilience, adaptability and transformability. *Ecology and Society, 15*(4), 20–28. https://doi.org/10.5751/ES-03610-150420.
Foodrescue.net. (2019). *K-12 Food Rescue* [online]. Available at: https://www.foodrescue.net/. Accessed 5 March 2019.
Foodtank. (2019). *15 organizations creating edible landscapes* [online]. Available at: https://foodtank.com/news/2018/07/organizations-creating-edible-landscapes/. Accessed 25 January 2019.
Forbes. (2018). *Rising CO_2 is reducing the nutritional value of our food* [online]. Available at: https://www.forbes.com/sites/fionamcmillan/2018/05/27/rising-co2-is-reducing-the-nutritional-value-of-our-food/#21cad9485133. Accessed 12 February 2019.
Forbes.com. (2018). *15 booming real estate markets that are trending in 2018* [online]. Available at: https://www.forbes.com/sites/forbesrealestatecouncil/2018/04/03/15-booming-real-estate-markets-that-are-trending-in-2018/#2db0add226f6. Accessed 20 December 2019.
Fortlauderdale.gov. (2018). *Vision plan: Fast forward Fort Lauderdale: Our city, our vision 2035* [online]. Available at: https://www.fortlauderdale.gov/departments/city-manager-s-office/structural-innovation-division/vision-plan. Accessed 16 October 2018.
Frayne, B., Pendleton, W., Crush, J., Acquah, B., Battersby-Lennard, J., Bras, E., et al. (2010). *The state of urban food insecurity in Southern Africa*. Urban Food Security Series No. 2. Kingston and Cape Town: Queen's University and AFSUN.
FRBSF.org. (2018). *Q&A: Climate adaptation and resilience from a community development perspective*. Available at: https://www.frbsf.org/community-development/blog/qa-climate-adaptation-and-resilience-from-a-community-development-perspective/?utm_source=verticalresponse&utm_medium=email&utm_campaign=climate-and-cd. Accessed 20 September 2018.
Fünfgeld, H., & McEvoy, D. (2012a). *Framing climate change adaptation in policy and practice* (Working Paper 1). Melbourne: Victorian Center for Climate Change Adaptation Research. Available at: http://vccar.org.au/files/vccar/Framing_project_workingpaper1_190411.pdf. Accessed 8 February 2012.
Fünfgeld, H., & McEvoy, D. (2012b). Resilience as a useful concept for climate change adaptation? In S. Davoudi, K. Shaw, L. J. Haider, A. E. Quinlan, G. D. Peterson, C. Wilkinson, H. Fünfgeld, D. McEvoy, & L. Porter (Eds.), Resilience: A bridging concept or a dead end? "Reframing" resilience: Challenges for planning theory and practice interacting traps: Resilience assessment of a pasture management system in Northern Afghanistan urban resilience: What does it mean in planning practice? Resilience as a useful

concept for climate change adaptation? The politics of resilience for planning: A cautionary note. *Planning Theory & Practice, 13*(2), 324–328. https://doi.org/10.1080/14649357.2012.677124.

Gimenez, C. (2018, October). Keynote given at the 10th Annual Southeast Florida Regional Climate Leadership Summit, Miami Beach, FL.

Gosliner, W., Brown, D., Sun, B. C., Woodward-Lopez, G., & Crawford, P. B. (2018). Availability, quality and price of produce in low-income neighbourhood food stores in California raise equity issues. *Public Health Nutrition, 21*(9), 1639–1648. https://doi.org/10.1017/S1368980018000058.

Greenbiz.com. (2018). *3 ways real estate developers can stay ahead of climate change* [online]. Available at: https://www.greenbiz.com/article/3-ways-real-estate-developers-can-stay-ahead-climate-change. Accessed 29 January 2018.

Gundersen, C. A., Dewey, A., Crumbaugh, M. K., & Engelhard, E. (2016). *Map the meal gap 2016: Food insecurity and child food insecurity estimates at the county level*. Chicago, IL: Feeding America.

Haider, L. J., Quinlan, A. E., & Peterson, G. D. (2012). Interacting traps: Resilience assessment of a pasture management system in Northern Afghanistan. In S. Davoudi, K. Shaw, L. J. Haider, A. E. Quinlan, G. D. Peterson, C. Wilkinson, H. Fünfgeld, D. McEvoy, & L. Porter (Eds.), Resilience: A bridging concept or a dead end? "Reframing" resilience: Challenges for planning theory and practice interacting traps: Resilience assessment of a pasture management system in Northern Afghanistan urban resilience: What does it mean in planning practice? Resilience as a useful concept for climate change adaptation? The politics of resilience for planning: A cautionary note. *Planning Theory & Practice, 13*(2), 324–328. https://doi.org/10.1080/14649357.2012.677124.

Harvard Health Publishing. (2015). *Nutritional psychiatry: Your brain on food* [online]. Available at: https://www.health.harvard.edu/blog/nutritional-psychiatry-your-brain-on-food-201511168626. Accessed 10 February 2019.

Harvard TH Chan School of Public Health. (2019). *Healthy eating plate vs. USDA's MyPlate* [online]. Available at: https://www.hsph.harvard.edu/nutritionsource/healthy-eating-plate-vs-usda-myplate/. Accessed 13 February 2019.

Health.gov. (2019). *Quick guide to health literacy* [online]. Available at: https://health.gov/communication/literacy/quickguide/factsbasic.htm#six. Accessed 23 May 2019.

Healthymiamidade.org. (2019a). *Home* [online]. Available at: https://www.healthymiamidade.org/. Accessed 20 February 2019.

Healthymiamidade.org. (2019b). *Health and the built environment* [online]. Available at: https://www.healthymiamidade.org/committees/health-and-the-built-environment/. Accessed 2 March 2019.

Hodgson, G. M. (2006). What are institutions? *Journal of Economic Issues, 40*(1), 1–25. https://doi.org/10.1177/0170840607067832.

Holling, C. S. (1973). Resilience and stability of ecological systems. *Annual Review of Ecology and Systematics*, 4, 1–23. https://doi.org/10.1146/annurev.es.04.110173.000245.

Holling, C. S. (1996). Engineering resilience versus ecological resilience. In P. Schulze (Ed.), *Engineering within ecological constraints*, p. 32. Washington, DC: The National Academies Press.

Holling, C. S., & Gunderson, L. H. (2002). Resilience and adaptive cycles. In L. H. Gunderson & C. S. Holling (Eds.), *Panarchy: Understanding transformations in human and natural systems* (pp. 25–62). Washington, DC: Island Press.

Houston, M. (2018, October 17). Personal interview.

Huntjens, P., Lebel, L., Pahl-Wostl, C., Camkin, J., Schulze, R., & Kranz, N. (2012). Institutional design propositions for the governance of adaptation to climate change in the water sector. *Global Environmental Change*, 22, 67–81. https://doi.org/10.1016/j.gloenvcha.2011.09.015.

ICIC.org. (2015a). *Resilient food systems, resilient cities: Recommendations for the city of Boston. Executive summary* [online]. Available at: http://icic.org/wp-content/uploads/2016/04/ICIC_Food_Systems_ExecutiveSummary_final.pdf. Accessed 15 January 2019.

ICIC.org. (2015b). *Resilient food systems, resilient cities: Recommendations for the city of Boston. Implementation roadmap* [online]. Available at: http://icic.org/wp-content/uploads/2016/04/ICIC_Food_Systems_Roadmap_final_v2.pdf. Accessed 15 January 2019.

ICMA.org. (2018). *Esri smart communities case study series: Ft. Lauderdale: Developing a resilient, smart city* [online]. Available at: https://icma.org/sites/default/files/308254_16-276%20Esri%20Case%20Study%20Ft%20Lauderdale-web.pdf. Accessed 15 October 2018.

IISD.org. (2018). *Resilience* [online]. Available at: https://www.iisd.org/program/resilience. Accessed 13 February 2019.

Insightcced.org. (2019). *The Color of Wealth in Miami* [online]. Available at: https://insightcced.org/the-color-of-wealth-in-miami/. Accessed 2 March 2019.

IPCC.ch. (2018). *Summary for policymakers of IPCC special report on global warming of 1.5°C approved by governments* [online]. Available at: https://www.ipcc.ch/2018/10/08/summary-for-policymakers-of-ipcc-special-report-on-global-warming-of-1-5c-approved-by-governments/. Accessed 16 December 2018.

IWRA. (2019). *New IWRA report on "Developing a global compendium on water quality guidelines"!* [online]. Available at: https://www.iwra.org/waterqualityreport/. Accessed 10 May 2019.

Jenkins, A., Keeffe, G., & Hall, N. (2015). Planning urban food production into today's cities. *Future of Foods: Journal on Food, Agriculture and Society*, 3, 35–47.

Jurado, J. (2018, October 17). Personal interview.

Kauffman.org. (2018). *The Kauffman index: Miami-Fort Lauderdale-Pompano Beach* [online]. Available at: https://www.kauffman.org/kauffman-index/profile?loc=33100&name=miami-fort-lauderdale-pompano-beach&breakdowns=growth|overall,startup-activity|overall,main-street|overall. Accessed 20 December 2018.

Kim, T. J., & von dem Knesebeck, O. (2018). Income and obesity: What is the direction of the relationship? A systematic review and meta-analysis. *BMJ Open, 8*(1), e019862. https://doi.org/10.1136/bmjopen-2017-019862.

Kinzig, A. P., Ryan, P., Etienne, M., Allison, H., Elmqvist, T., & Walker, B. H. (2006). Resilience and regime shifts: Assessing cascading effects. *Ecology and Society, 11*(1), 20.

Korenman, S., Miller, J. E., & Siaastad, J. E. (1995). Long-term poverty and child development in the United States: Results from the NLSY. *Children and Youth Services Review, 17*(1–2), 127–155. https://doi.org/10.1016/0190-7409(95)00006-x.

Lacity.org. (2017). *City council passes zero waste LA program* [online]. Available at: https://www.lacity.org/blog/city-council-passes-zero-waste-la-program. Accessed 2 March 2019.

LaPradd, C. (2018, November 2). Personal interview.

Letson, D. (2017). Climate change and food security: Florida's agriculture in the coming decades. In A. Schmitz, P. L. Kennedy, & T. G. Schmitz (Eds.), *World agricultural resources and food security* (Frontiers of Economics and Globalization, Vol. 17, pp. 85–102). Bingley: Emerald Publishing. https://doi.org/10.1108/s1574-871520170000017007.

Little, R. G. (2002). Controlling cascading failure: Understanding the vulnerabilities of interconnected infrastructures. *Journal of Urban Technology, 9*(1), 109–123. https://doi.org/10.1080/106307302317379855.

Longstaff, P. H., Armstrong, N. J., Perrin, K., Parker, W. M., & Hidek, M. A. (2010). Building resilient communities: A preliminary framework for assessment. *Homeland Security Affairs, VI*(3), 1–23.

Los Angeles Food Policy Council. (2019). *Join the good food movement* [online]. Available at: https://www.goodfoodla.org/. Accessed 13 February 2019.

Lowi, T. J. (1979). *The end of liberalism: The second republic in the United States* (2nd ed.). New York: W. W. Norton.

MacDonald, J. M., & Nelson, P. E. (1991). Do the poor still pay more? Food price variations in large metropolitan areas. *Journal of Urban Economics, 30*, 344–359.

Mattiuzzi, E. (2018). *Q&A: Climate adaptation and resilience from a community development perspective.* Federal Reserve Bank of San Francisco [online]. Available at: https://www.frbsf.org/community-development/blog/qa-climate-adaptation-and-resilience-from-a-community-development-perspective/. Accessed 5 December 2018.

Mayor's Office of Resilience & Racial Equity. (2017). *Resilient Boston: An equitable and connected city* [online]. Available at: http://www.100resilientcities.org/wp-content/uploads/2017/07/Boston-Resilience-Strategy-Reduced-PDF.pdf. Accessed 14 January 2019.

Mdc.maps.arcgis.com. (2018). *What is Miami-Dade County doing about sea level rise?* [online]. Available at: https://mdc.maps.arcgis.com/apps/Shortlist/index.html?appid=51003eca3778442ca5b8bc8c0868920a&mc_cid=d589361308&mc_eid=5e441ffbf9. Accessed 20 September 2018.

Miami Herald. (2017). *Miami gets $200 million to spend on sea rise as voters pass Miami Forever Bond* [online]. Available at: http://www.miamiherald.com/news/politics-government/election/article183336291.html. Accessed 29 January 2018.

Miami Herald. (2018a). *'Why are we the guinea pig?': Climate change project divides a Miami Beach neighborhood* [online]. Available at: https://www.miamiherald.com/article223054220.html. Accessed 2 December 2018.

Miami Herald. (2018b). *Your flood insurance premium is going up again, and that's only the beginning* [online]. Available at: https://www.miamiherald.com/news/state/florida/article215162440.html. Accessed 4 September 2018.

Miami Herald. (2018c). *Climate gentrification: Is sea rise turning Miami high ground into a hot commodity?* [online]. Available at: https://www.miamiherald.com/news/local/environment/article222547640.html. Accessed 3 January 2019.

Miami Herald. (2018d). *Kids are suing Gov. Rick Scott to force Florida to take action on climate change* [online]. Available at: https://www.miamiherald.com/news/local/environment/article208967284.html. Accessed 10 December 2018.

Miami Herald. (2018e). *It's really hot in Miami, but the feds don't require A/C in public housing* [online]. Available at: https://www.miamiherald.com/news/local/community/miami-dade/edison-liberty-city/article217030505.html. Accessed 30 October 2018.

Miami Herald. (2018f). *Florida heat is already hard on outdoor workers: Climate change will raise health risks* [online]. Available at: https://www.miamiherald.com/news/local/environment/article220833555.html. Accessed 20 December 2018.

Miami Herald. (2018g). *Curbelo considering 2020 Miami-Dade mayoral bid* [online]. Available at: https://www.miamiherald.com/news/politics-government/article222319520.html. Accessed 30 November 2018.

Miami Herald. (2018h). *How will sea level rise affect your home? Miami is creating a tool that will show you* [online]. Available at: http://www.miamiherald.com/news/local/community/miami-dade/article209710464.html. Accessed 4 May 2018.

Miami Herald. (2018i). *Florida leads nation in property at risk from climate change* [online]. Available at: https://www.miamiherald.com/news/local/environment/article29029159.html. Accessed 12 April 2019.

Miami Herald. (2019a). *DeSantis announces sweeping fixes meant to clean up Florida water woes* [online]. Available at: https://www.miamiherald.com/news/local/environment/article224219365.html. Accessed 30 January 2019.

Miami Herald. (2019b). *Threat to Miami economy from stock trouble, shutdown— And robots—Is limited, experts say* [online]. Available at: https://www.miamiherald.com/news/business/article224623860.html. Accessed 10 February 2019.

Miami Herald. (2019c). *A $3 billion problem: Miami-Dade's septic tanks are already failing due to sea rise* [online]. Available at: https://www.miamiherald.com/news/local/environment/article224132115.html. Accessed 15 January 2019.

Miami Housing Solutions Lab. (2019). *Miami housing solutions lab* [online]. Available at: http://cdn.miami.edu/wda/cce/Documents/Miami-Housing-Solutions-Lab/index.html. Accessed 19 January 2019.

Miami New Times. (2018). *Sea-level rise leading to Florida mangrove "death march" study warns* [online]. Available at: http://www.miaminewtimes.com/news/sea-level-rise-killing-florida-mangroves-fiu-study-warns-10315556. Accessed 7 May 2018.

Miami21.org. (2018). *Project vision* [online]. Available at: http://www.miami21.org/. Accessed 12 October 2018.

Miami.curbed.com. (2017). *Miami is the country's 4th most valuable housing market, per report* [online]. Available at: https://miami.curbed.com/2017/12/29/16829424/miami-housing-market-report-growth. Accessed 29 January 2018.

Miamibeachfl.gov. (2018). *Sustainability plan: Energy economic zone plan* [online]. Available at: https://www.miamibeachfl.gov/wp-content/uploads/2017/12/City-of-Miami-Beach-Sustainabilty-Plan_FINAL.pdf. Accessed 22 October 2018.

Miamiclimatealliance.org. (2019). *Home* [online]. Available at: http://miamiclimatealliance.org/. Accessed 10 February 2019.

Miamifoundation.org. (2019). *About us: Advancing quality of life in Greater Miami* [online]. Available at: https://miamifoundation.org/about/. Accessed 10 January 2019.

Miamigov.com. (2019). *Coastal and stormwater infrastructure* [online]. Available at: https://www.miamigov.com/Government/MiamiClimateSolutions/Coastal-and-Stormwater-Infrastructure. Accessed 2 March 2019.

Milanurbanfoodpolicypact.org. (2017a). *Milan urban food policy pact* [online]. Available at: http://www.milanurbanfoodpolicypact.org/. Accessed 2 February 2019.

Milanurbanfoodpolicypact.org. (2017b). *Monitoring framework* [online]. Available at: https://www.milanurbanfoodpolicypact.org/monitoring-framework/. Accessed 22 January 2018.

Milly, P. C. D., Betancourt, J., Falkenmark, M., Hirsch, R. M., Kundzewicz, Z. W., Lettenmaier, D. P., et al. (2008). Stationarity is dead: Whither water management? *Science, 319*(5863), 573–574. https://doi.org/10.1126/science.1151915.

MindBodyGreen.com. (2018). *40 million Americans don't have access to enough food: These companies are trying to change that* [online]. Available at: https://www.mindbodygreen.com/articles/naked-juice-wholesome-wave-pepsi-food-access-initiatives. Accessed 12 February 2019.

Mohl, R. A. (2001). Whitening Miami: Race, housing, and government policy in twentieth-century Dade County. In *The Florida historical quarterly, vol 79, no. 3: In reconsidering race relations in early twentieth-century Florida (Winter)* (pp 319–345). Florida: Historical Society.

Moodys.com. (2017). *Announcement: Moody's: Climate change is forecast to heighten US exposure to economic loss placing short- and long-term credit pressure on US states and local governments* [online]. Available at: https://www.moodys.com/research/Moodys-Climate-change-is-forecast-to-heighten-US-exposure-to--PR_376056. Accessed 29 January 2018.

Moser, C., & Satterthwaite, D. (2010). Towards pro-poor adaptation to climate change in the urban centers of low- and middle-income countries. In R. Mearns & A. Norton (Eds.), *Social dimensions of climate change: Equity and vulnerability in a warming world* (pp. 231–258). Washington, DC: World Bank.

Myboca.us. (2018). *Media release: The city of Boca Raton hires first sustainability manager* [online]. Available at: https://www.myboca.us/DocumentCenter/View/17314/Media-Release---City-of-Boca-Raton-Hires-First-Sustainability-Manager-05312018-PDF. Accessed 16 October 2018.

NCA2018.Globalchange.gov. (2019). *Fourth national climate assessment* [online]. Available at: https://nca2018.globalchange.gov. Accessed 1 February 2019.

Nesheim, M. C., Oria, M., & Tsay Yih, P. (Eds.). (2015). *A framework for assessing effects of the food system*. Washington, DC: National Academic Press.

New Orleans Food Policy Advisory Committee. (2019). *Who we are* [online]. Available at: http://www.nolafoodpolicy.org/index.php?page=who-we-are. Accessed 13 February 2019.

Newsweek. (2014). *A town called Malnourished* [online]. Available at: https://www.newsweek.com/2014/04/11/town-called-malnourished-248087.html. Accessed 3 February 2019.

Norris, F. H., Stevens, S. P., Pfefferbaum, B., Wyche, K. F., & Pfefferbaum, R. L. (2008). Community resilience as a metaphor, theory, set of capabilities, and strategy for disaster readiness. *American Journal of Community Psychology, 41*, 127–150.

NYC Food Policy. (2019a). *Welcome* [online]. Available at: https://www1.nyc.gov/site/foodpolicy/index.page. Accessed 13 February 2019.

NYC Food Policy. (2019b). *About* [online]. Available at: https://www1.nyc.gov/site/foodpolicy/about/nyc-food-policy.page. Accessed 13 February 2019.

NYC Food Policy. (2019c). *Food metrics report* [online]. Available at: https://www1.nyc.gov/site/foodpolicy/about/food-metrics-report.page. Accessed 13 February 2019.

Nytimes.com. (2018). *Major climate report describes a strong risk of crisis as early as 2040* [online]. Available at: https://www.nytimes.com/2018/10/07/climate/ipcc-climate-report-2040.html. Accessed 20 December 2018.

Olsson, P., Gunderson, L. H., Carpenter, S., Ryan, P., Lebel, L., Folke, C., et al. (2006). Shooting the rapids: Navigating transitions to adaptive governance of social-ecological systems. *Ecology and Society, 11*(1), 18.

Palm Beach County. (2015, October). *Hunger relief plan Palm Beach County* [online]. Available at: http://discover.pbcgov.org/communityservices/humanservices/PDF/News/Palm_Beach_FRAC_100515_Edition-v2.pdf. Accessed 25 January 2019.

Pelling, M. (2003). *The vulnerability of cities: Natural disasters and social resilience*. London: Earthscan.

Porter, L., & Davoudi, S. (2012). The politics of resilience for planning: A cautionary note. In S. Davoudi, K. Shaw, L. J. Haider, A. E. Quinlan, A. E., G. D. Peterson, C. Wilkinson, H. Fünfgeld, D. McEvoy, & L. Porter (Eds.), Resilience: A bridging concept or a dead end? "Reframing" resilience: Challenges for planning theory and practice interacting traps: Resilience assessment of a pasture management system in Northern Afghanistan urban resilience: What does it mean in planning practice? Resilience as a useful concept for climate change adaptation? The politics of resilience for planning: A cautionary note. *Planning Theory & Practice, 13*(2), 324–328. https://doi.org/10.1080/14649357.2012.677124.

Porter, J. R., et al. (2014). Food security and food production systems. In *Climate change 2014: Impacts, adaptation, and vulnerability. Part A: Global and sectoral aspects. Contribution of Working Group II to the Fifth Assessment Report of the Intergovernmental Panel on Climate Change* (pp. 485–533).

Quinlan, A. (2003). Resilience and adaptive capacity: Key components of a sustainable social-ecological system. *IHDP Update, 2*, 4–5.

Radical Partners. (2019). *100 great ideas* [online]. Available at: https://static1.squarespace.com/static/56526d51e4b083936d517729/t/5ca4d099eb3931713a53a29b/1554305178677/100+Great+Ideas+Climate+Resilience+%26+Sustainability+Final+Report.pdf. Accessed 10 May 2019.

Resalliance.org. (2018). *Resilience* [online]. Available at: https://www.resalliance.org/resilience. Accessed 16 December 2018.

Resilience Alliance. (2007). *Assessing and managing resilience in social-ecological systems: A practitioner's workbook (resilience alliance)*. Available at: www.resalliance.org.

Resilience Alliance. (2010). *Assessing resilience in social-ecological systems: A workbook for practitioners, version 2.0* [online]. Available at: http://www.resalliance.org/3871.php. Accessed 27 February 2018.

Resilient305. (2019). *Resilient Greater Miami & the beaches*. Miami, FL.

Resilient305.org. (2018). *Resilient Greater Miami & the beaches: Preliminary resilience assessment #Resilient305* [online]. Available at: http://resilient305.com/assets/pdf/170905_GM&B%20PRA_v01-2.pdf. Accessed 29 January 2018.

Ridzuan, A. A., Oktari, R. S., Zainol, N. A. M., Abdullah, H., Liaw, J. O. H., Mohaiyadin, N. M. H., et al. (2018). *Community resilience elements and community risk perception at Banda Aceh province, Aceh, Indonesia*. MATEC Web of Conferences (Vol. 229, p. 01005). EDP Sciences.

RRAC. (2009). *Building resilient communities, from ideas to sustainable action*. London: Risk and Regulation Advisory Council.

Sage, C. (2014). The transition movement and food sovereignty: From local resilience to global engagement in food system transformation. *Journal of Consumer Culture, 14*(2), 254–275. https://doi.org/10.1177/1469540514526281.

Satterthwaite, D., Dodman, D., & Bicknell, J. (2009). Conclusions: Local development and adaptation. In J. Bicknell, D. Dodman, & D. Satterthwaite (Eds.), *Adapting cities to climate change: Understanding and addressing the development challenges* (pp. 359–383). London: Earthscan.

Scharmer, O. (2019). *Addressing the blind spot of our time: An executive summary of the new book by Otto Scharmer Theory U: Leading from the future as it emerges* [online]. Available at: https://www.presencing.org/assets/images/theory-u/Theory_U_Exec_Summary.pdf. Accessed 13 February 2019.

Scheffer, M. (2009). *Critical transitions in nature and society*. Princeton, NJ: Princeton University Press.

Sciencemarchmiami.org. (2019). *Who we are* [online]. Available at: https://www.sciencemarchmiami.org/. Accessed 10 May 2019.

Seafoodsource. (2017). *Atlantic Sapphire building USD 350 million land-based salmon farm in Miami* [online]. Available at: https://www.seafoodsource.com/news/aquaculture/atlantic-sapphire-building-usd-350-million-land-based-salmon-farm-in-miami. Accessed 2 March 2019.

Seville, E. (2009). Resilience: Great concept…but what does it mean for organisations? In Ministry of Civil Defence & Emergency Management (Ed.), *Community resilience: Research, planning and civil defence emergency management*. Wellington, NZ: Ministry of Civil Defence & Emergency Management.

Shaw, K. (2012). "Reframing" resilience: Challenges for planning theory and practice. In S. Davoudi & L. Porter (Eds.), Applying the resilience perspective to planning: Critical thoughts from theory and practice. *Planning Theory & Practice, 13*(2), 308–312. https://doi.org/10.1080/14649357.2012.677124.

Shaw, K., & Maythorne, L. (2012). Managing for local resilience: Towards a strategic approach. *Public Policy and Administration, 28*(1), 43–65.

Singleton, C. R., Sen, B., & Affuso, O. (2015). Disparities in the availability of farmers markets in the United States. *Environmental Justice (Print), 8*(4), 135–143. https://doi.org/10.1089/env.2015.0011.

Socialventurepartners.org. (2019). *Social Venture Partners Miami: A global venture philanthropy model arrives* [online]. Available at: http://www.socialventurepartners.org/wp-content/uploads/2017/01/Miami-Overview-Interest-Form.pdf. Accessed 10 February 2019.

Southeastfloridaclimatecompact.org. (2015). *Unified sea level rise projection.* Southeast Florida [online]. Available at: http://www.southeastfloridaclimatecompact.org/wp-content/uploads/2015/10/2015-Compact-Unified-Sea-Level-Rise-Projection.pdf. Accessed 2 March 2019.

Southeastfloridaclimatecompact.org. (2017). *Advancing resilience solutions through regional action* [online]. Available at: http://www.southeastfloridaclimatecompact.org/. Accessed 18 December 2017.

Southeastfloridaclimatecompact.org. (2018a). *Agriculture* [online]. Available at: http://www.southeastfloridaclimatecompact.org/recommendation-category/ag/. Accessed 20 September 2018.

Southeastfloridaclimatecompact.org. (2018b). *About: What is the compact?* [online]. Available at: http://www.southeastfloridaclimatecompact.org/. Accessed 18 December 2017.

Southeastfloridaclimatecompact.org. (2018c). *1 page flyer* [online]. Available at: http://www.southeastfloridaclimatecompact.org/wp-content/uploads/2017/12/compact-1-page-flyer-ia-final-sa.pdf. Accessed 18 December 2017.

Southeastfloridaclimatecompact.org. (2018d). *Explore the RCAP* [online]. Available at: http://www.southeastfloridaclimatecompact.org/news/can-solve-climate-change-without-business-community-south-florida-says-no/. Accessed 23 December 2017.

Southeastfloridaclimatecompact.org. (2018e). *Can you solve climate change without the business community? South Florida says no* [online]. Available at: http://www.southeastfloridaclimatecompact.org/recommendations/. Accessed 20 December 2018.

Southeastfloridaclimatecompact.org. (2019a). *Public health* [online]. Available at: http://www.southeastfloridaclimatecompact.org/recommendation-category/ph/. Accessed 5 January 2019.

Southeastfloridaclimatecompact.org. (2019b). *Agriculture* [online]. Available at: http://www.southeastfloridaclimatecompact.org/recommendation-category/ag/. Accessed 5 January 2019.

Southmiamifl.gov. (2018). *Going green!* [online]. Available at: https://www.southmiamifl.gov/512/Going-Green. Accessed 20 September 2018.

Squire, T. (2018, November 19). Personal interview.

Stanford Social Innovation Review. (2011). *Collective impact* [online]. Available at: https://ssir.org/articles/entry/collective_impact. Accessed 25 January 2019.

Strollingtheheifers.com. (2018). *Locavore index: How locavore is your state?* [online]. Available at: https://www.strollingoftheheifers.com/locavore/. Accessed 20 February 2019.

Sun-Sentinel.com. (2016). *Florida lawmakers consider low-cost grocery store loans to promote healthy eating* [online]. Available at: http://www.sun-sentinel.com/health/fl-food-desert-supermarket-legislation-20160205-story.html. Accessed 29 January 2018.

Swanstrom, T. (2008). *Regional resilience: A critical examination of the ecological framework* (Working Paper, 2008[7]). Berkeley: University of California, Institute of Urban and Regional Development.

Tampa Bay Times. (2019). *Farm to fable* [online]. Available at: http://www.tampabay.com/projects/2016/food/farm-to-fable/. Accessed 13 February 2019.

Tecco, N., Coppola, F., Sottile, F., & Peano, C. (2017). Urban gardens and institutional fences. *Future of Food: Journal on Food, Agriculture and Society, 5*(1), 70–78.

The CLEO Institute. (2019). *Empowering capable climate communicators symposium* [online]. Available at: https://www.cleoinstitute.org/symposium-2019. Accessed 10 May 2019.

The Florida Climate Pledge. (2018). *The pledge* [online]. Available at: https://floridaclimatepledge.org/. Accessed 27 June 2018.

The Food Trust. (2012). *A healthier future for Miami-Dade County: Expanding supermarket access in areas of need* [online]. Available at: http://thefoodtrust.org/uploads/media_items/miami-dade-supermarket-report.original.pdf. Accessed 26 January 2019.

The Hill. (2018). *Poll: Record number of Americans believe in man-made climate change* [online]. Available at: https://thehill.com/policy/energy-environment/396487-poll-record-number-of-americans-believe-in-man-made-climate-change. Accessed 2 February 2019.

The Invading Sea. (2019). *About us* [online]. Available at: https://www.theinvadingsea.com/about-us/. Accessed 10 February 2019.

The National Wildlife Federation. (2019). *The Everglades* [online]. Available at: https://www.nwf.org/Educational-Resources/Wildlife-Guide/Wild-Places/Everglades. Accessed 2 March 2019.

The Nature Conservancy. (2018). *Climate action from the ground up* [online]. Available at: https://global.nature.org/content/climate-action-from-the-ground-up?src3=e.gp.nat.Sept2018.National.readmore. Accessed 20 September 2018.

The New Tropic. (2015). *How tourism impacts Miami* [online]. Available at: https://thenewtropic.com/tourism-economy-culture/. Accessed 10 December 2018.

The New Tropic. (2019). *About us* [online]. Available at: https://thenewtropic.com/about/. Accessed 10 May 2019.

The Palm Beach Post. (2016). *Palm Beach County agriculture touted as economic engine at summit* [online]. Available at: https://www.palmbeachpost.com/article/20160504/BUSINESS/812068300. Accessed 2 March 2019.

The Wall Street Journal. (2018). *Rising sea levels reshape Miami's housing market* [online]. Available at: https://www.wsj.com/articles/climate-fears-reshape-miamis-housing-market-1524225600?. Accessed 4 May 2018.

The Washington Post. (2017). *Junk food is cheap and healthful food is expensive, but don't blame the farm bill* [online]. Available at: https://www.washingtonpost.com/lifestyle/food/im-a-fan-of-michael-pollan-but-on-one-food-policy-argument-hes-wrong/2017/12/04/c71881ca-d6cd-11e7-b62d-d9345ced896d_story.html?utm_term=.467a9e106b47. Accessed 10 May 2019.

Thepatchgarden.com. (2019). *About* [online]. Available at: http://thepatchgarden.com/about. Accessed 10 February 2019.

Theunderline.org. (2019). *Home* [online]. Available at: https://www.theunderline.org/. Accessed 13 February 2019.

Torriente, S. (2018, October 24). Personal interview.

Tulane University. (2019). *Food deserts in America (infographic)* [online]. Available at: https://socialwork.tulane.edu/blog/food-deserts-in-america. Accessed 2 February 2019.

Tyler, S. R., Keller, M., & Swanson, D. (2013). *Climate resilience and food security: A framework for planning and monitoring*. Winnipeg: The International Institute for Sustainable Development.

Tyler, S., & Moench, M. (2012). A framework for urban climate resilience. *Climate and Development, 4*(4), 311–326. https://doi.org/10.1080/17565 529.2012.745389.

Tyler, S., & Reed, S. O. (2011). Results of resilience planning. In M. Moench, S. Tyler, & J. Lage (Eds.), *Catalyzing urban climate resilience: Applying resilience concepts to planning practice in the ACCCRN program 2009–2011* (pp. 239–270). Boulder, CO: ISET.

UN Org. (2014). *World's population increasingly urban with more than half living in urban areas* [online]. Available at: http://www.un.org/en/development/desa/news/population/world-urbanization-prospects-2014.html. Accessed 22 January 2018.

UN Org. (2017). *UN, partners warn 108 million people face severe food insecurity worldwide* [online]. Available at: http://www.un.org/apps/news/story.asp?NewsID=56472#.WmZFu5M-cWo. Accessed 22 January 2018.

UNFCC.int. (2019). *What is the Paris Agreement?* [online]. Available at: https://unfccc.int/process-and-meetings/the-paris-agreement/what-is-the-paris-agreement. Accessed 3 January 2019.

United States Department of Labor Bureau of Labor Statistics. (2019). *Southeast Information Office* [online]. Available at: https://www.bls.gov/regions/southeast/news-release/consumerexpenditures_miami.htm. Accessed 20 January 2019.

United States Environmental Protection Agency. (2019). *Sustainable management of food basics* [online]. Available at: https://www.epa.gov/sustainable-management-food/sustainable-management-food-basics. Accessed 2 February 2019.

UnitedWayMiami.org. (2017). *The 2017 united way ALICE report* [online]. Available at: https://unitedwaymiami.org/wp-content/uploads/2014/11/10351-EXT-ALICE-2017-FINAL-single-pgs-1.pdf. Accessed 4 January 2019.

University of Miami. (2019a). *Welcome* [online]. Available at: http://www.miami.edu/civic. Accessed 10 February 2019.

University of Miami. (2019b). *Sustainability* [online]. Available at: https://greenu.miami.edu/topics/food-and-well-being/community-gardens/index.html. Accessed 10 February 2019.

University of Notre Dame. (2019). *Urban adaptation assessment: Better data for planning for your city's future* [online]. Available at: https://gain-uaa.nd.edu/. Accessed 23 May 2019.

Urbangreenworks.org. (2019). *Urban GreenWorks, restoring the economic, physical and social health of under-served communities* [online]. Available at: https://www.urbangreenworks.org/. Accessed 10 February 2019.

Urbanland.uli.org. (2018). *Living with rising sea levels: Miami Beach's plans for resilience* [online]. Available at: https://urbanland.uli.org/sustainability/living-rising-sea-levels-miami-beachs-plans-resilience/. Accessed 22 October 2018.

US Army Corps of Engineers. (2019). *Miami-Dade Back Bay coastal storm risk management feasibility study* [online]. Available at: https://www.saj.usace.army.mil/MiamiDadeBackBayCSRMFeasibilityStudy/. Accessed 2 February 2019.

USDA. (2010). *Access to affordable, nutritious food is limited in "Food Deserts"* [online]. Available at: https://www.ers.usda.gov/amber-waves/2010/march/access-to-affordable-nutritious-food-is-limited-in-food-deserts/. Accessed 20 December 2018.

USDA Economic Research Service. (2019). *Food environment Atlas* [online]. Available at: https://www.ers.usda.gov/data-products/food-environment-atlas/. Accessed 5 February 2019.
USDA NIFA. (2019). *Obesity* [online]. Available at: https://nifa.usda.gov/topic/obesity. Accessed 2 February 2019.
Vale, L. J., & Campanella, T. J. (2005). *The resilient city: How modern cities recover from disaster.* New York: Oxford University Press.
Vildosola, D. (2019). *Community gardening in the first step to sustainable living* [online]. Available at: https://umiami.maps.arcgis.com/apps/MapJournal/index.html?appid=88c1fbb1c06d441298a4ac2fb2574333. Accessed 10 February 2019.
Walker, B., Holling, C. S., Carpenter, S., & Kinzig, A. (2004). Resilience, adaptability and transformability in social-ecological systems. *Ecology and Society, 9*(2), 5.
Wardekker, J. A., de Jong, A., Knopp, J. M., & van der Sluijs, J. P. (2010). Operationalizing a resilience approach to adapting a delta to uncertain climate changes. *Technological Forecasting and Social Change, 77*, 987–998. https://doi.org/10.1016/j.techfore.2009.11.005.
Watts, N., Amann, M., Ayeb-Karlsson, S., Belesova, K., Bouley, T., Boykoff, M., et al. (2017). The Lancet countdown on health and climate change: from 25 years of inaction to a global transformation for public health. *Lancet, 391*(10120), 581–630. https://doi.org/10.1016/S0140-6736(17)32464-9.
Whalen, R., & Zeuli, K. (2017). *Resilient cities require resilient food systems.* The Rockefeller Foundation [online]. Available at: https://www.rockefellerfoundation.org/blog/resilient-cities-require-resilient-food-systems/. Accessed 18 December 2017.
Whitehouse.gov. (2018). *OMB Bulletin No. 18-03* [online]. Available at: https://www.whitehouse.gov/wp-content/uploads/2018/04/OMB-BULLETIN-NO.-18-03-Final.pdf. Accessed 2 February 2019.
WHO.int. (2018a). *Obesity and overweight: Key facts* [online]. Available at: https://www.who.int/news-room/fact-sheets/detail/obesity-and-overweight. Accessed 16 December 2018.
WHO.int (2018b). *Public spending on health: A closer look at global trends* [online]. Available at: https://www.who.int/health_financing/documents/health-expenditure-report-2018/en/. Accessed 16 December 2018.
Wholesomewave.org. (2019). *Changing the world through food* [online]. Available at: https://www.wholesomewave.org/. Accessed 12 February 2019.
Wilkinson, C. (2012). Urban resilience: What does it mean in planning practice? In S. Davoudi, K. Shaw, L. J. Haider, A. E. Quinlan, G. D. Peterson, C. Wilkinson, H. Fünfgeld, D. McEvoy, L. Porter, & L. Porter (Eds.), Resilience: A bridging concept or a dead end? "Reframing" resilience:

Challenges for planning theory and practice interacting traps: Resilience assessment of a pasture management system in Northern Afghanistan urban resilience: What does it mean in planning practice? Resilience as a useful concept for climate change adaptation? The politics of resilience for planning: A cautionary note. *Planning Theory & Practice, 13*(2), 324–328. https://doi.org/10.1080/14649357.2012.677124.

Wilkinson, C., Porter, L., & Colding, J. (2010). Metropolitan planning and resilience thinking: A practitioner's perspective. *Critical Planning, 17*(17), 25–44.

Wilson, A. D. (2012). Beyond alternative: Exploring the potential for autonomous food spaces. *Antipode, 45*(3), 719–737. https://doi.org/10.1111/j.1467-8330.2012.01020.x.

Withanachchi, S., Kunchulia, I., Ghambashidze, G., Al Sidawi, R., Urushadze, T., & Ploeger, A. (2018). Farmers' perception of water quality and risks in the Mashavera River Basin, Georgia: Analyzing the vulnerability of the social-ecological system through community perceptions. *Sustainability, 10*(9), 3062.

Wlrn.org. (2017). *After hurricane Irma, food insecurity in Miami-Dade's poorest communities* [online]. Available at: http://wlrn.org/post/after-hurricane-irma-food-insecurity-miami-dades-poorest-communities. Accessed 29 January 2018.

Wlrn.org. (2018a). *Community groups begin work on hurricane plans for low-income neighborhoods in Miami-Dade, Broward* [online]. Available at: http://wlrn.org/post/community-groups-begin-work-hurricane-plans-low-income-neighborhoods-miami-dade-broward. Accessed 7 May 2018.

Wlrn.org. (2018b). *Coalition of environment groups launches push for state action on climate change* [online]. Available at: http://wlrn.org/post/coalition-environment-groups-launches-push-state-action-climate-change. Accessed 7 May 2018.

Wlrn.org. (2018c). *How trees can make South Florida more resilient against rising temperatures* [online]. Available at: http://wlrn.org/post/how-trees-can-make-south-florida-more-resilient-against-rising-temperatures. Accessed 7 May 2018.

Wlrn.org. (2018d). *Monday was a big day in climate and economic news. Here's what South Floridians should know* [online]. Available at: http://www.wlrn.org/post/monday-was-big-day-climate-and-economic-news-heres-what-south-floridians-should-know. Accessed 10 February 2019.

Wlrn.org. (2018e). *Temperatures in Florida are rising: For vulnerable patients, the heat can be life-threatening* [online]. Available at: http://www.wlrn.org/post/temperatures-florida-are-rising-vulnerable-patients-heat-can-be-life-threatening. Accessed 20 December 2018.

WPB.org. (2018). *Sustainability overview* [online]. Available at: http://wpb.org/Departments/Sustainability/Overview. Accessed 16 October 2018.

WTO. (2019). *Nutrition: Micronutrient deficiencies* [online]. Available at: https://www.who.int/nutrition/topics/vad/en/. Accessed 2 February 2019.

Wyncstations.org. (2018a). *The Takeaway: In face of looming challenges, building a resilient Miami* [online]. Available at: https://www.wnycstudios.org/story/mayor-francis-suarez-his-first-term-office?mc_cid=a44813592e&mc_eid=5e441ffbf9. Accessed 20 September 2018.

Wyncstations.org. (2018b). *The Takeaway: In Miami, combating climate change block by block* [online]. Available at: https://www.wnycstudios.org/story/combating-climate-change-local-level-miami?mc_cid=a44813592e&mc_eid=5e441ffbf9. Accessed 20 September 2018.

Zeuli, K., & Nijhuis, A. (2017). *The resilience of America's urban food systems: Evidence from five cities* [ebook]. Roxbury, MA: ICIC. Available at: http://icic.org/wp-content/uploads/2017/01/Rockefeller_ResilientFoodSystems_FINAL_post.pdf?x96880. Accessed 18 December 2017.

Zhu, C., Kobayashi, K., Loladze, I., Zhu, J., Jiang, Q., Xu, X., et al. (2018). Carbon dioxide (CO_2) levels this century will alter the protein, micronutrients, and vitamin content of rice grains with potential health consequences for the poorest rice-dependent countries. *Science Advances, 4*(5), eaaq1012. https://doi.org/10.1126/sciadv.aaq1012.

Ziervogel, G., & Ericksen, P. J. (2010). *Adapting to climate change to sustain food security: Advanced review* [online]. Available at: wires.wiley.com/climatechange.

Ziervogel, G., & Frayne, B. (2011). *Climate change and food security in Southern African cities* (Urban Food Security Series No. 8). Kingston and Cape Town: Queen's University and AFSUN.

Index

A
ABC Local 10 News, 59, 67
Academia, 103, 119, 152
Access to resources, 12
Adaptability, 4–9
Adaptive capacity, 39, 51, 68, 130, 138
Adefris, Zelalem, 150
Aedes mosquito, 50
Affordable housing, 76, 83, 86, 103, 112
African American, 44, 53, 59, 89
Aging population, 3, 28, 39
Agriculture, 25, 62, 130
Agrochemical, 19
Aldi, 64
ALICE report, 56
Allapattah, 54, 60
Allergy, 50
Amazon, 65
American Immigration Council, 41–43, 51
Recovery and Reinvestment Act, 40
American Medical Association (AMA), 50
Anthropogenic Ghg Emission, 25
Aquaculture, 64
Aquaponics, 142
Aquifer, 63, 84
Arditi-Rocha, Yoca, 101
Aridity, 25
Artificial Intelligence (AI), 3
Asthma, 50, 81

B
Baltimore, 116, 128, 132
Baltimore Food Policy Initiative (BFPI), 116
Baptist Health, 67
System, 104

Beach Erosion, 76
Beachside Montessori Village, 93–94
Beacon Council, 48, 76, 81, 103, 137, 151
Beeheavenfarm.com, 66
Bibbins, Shamar, 153
Bill De Blasio, 116
Bill Emerson Act, 93
Biogas, 116, 122
Biscayne Bay, 143
  flood, 88
Biz Bash, 94
Black residents, 44, 65
Bloomberg, 49–50
Blue Bin, 151
Blue-Green Algae, 41
Blue-Green Algae Task Force, 41
Boca Raton, 97
Body Mass Index (BMI), 20
Boston, 112–115
Brennan, Michael, 123
BRIDGe Corps, 104
Bridges, 26, 135, 144
Bristol City, 122
  Council, 122
Bristol Green Capital, 122
Broward County, 90–93
Brundtland, 7

C
California, 25, 38, 42, 63, 154
Canada, 42
Cancer, 2, 20
Carbon, 25, 77
  dioxide of, 25, 27, 93
  emission, 23
  tax, 78
Castro, Chris, 119
Castro, Fidel, 43
Cdc.gov, 18, 50

CDC REACH, 90
CGIAR, 25
Chief Resilience Officer (CRO), 80, 82, 84, 90, 97
Chief Science Officer, 41
Civic
  and Community Engagement (CCE), 103
  engagement, 105
Cleo Institute, 41, 101, 106
Climate
  adaptation, 14–17, 40, 78, 96, 101, 128, 21–23
  change, 3, 11, 15, 23, 38; action plan, 94; food accessibility, 21
  gentrification, 54
CNN.com, 43
Coconut, 150
Colombia, 42
Colon, 2, 20
Communication, 10, 13, 38, 87, 139, 143
Community
  activism, 99, 105, 155
  development, 43
  education, 149
  engagement, 98
  farm, 119
  knowledge, 13
  Nutrition Resilience, 29, 66, 69, 85, 98–100
  Supported Agriculture (CSA), 65, 144, 146
Conservation Law Foundation, 123
Consortium for a Healthier Miami-Dade, 86
Corporate Sustainability Network (CSN), 92
Cuba, 42–43
Cultural relevance, 29, 66, 85, 145, 150
Curbelo, Carlos, 78

D

Dania Beach, 93
Datausa.io, 56
Davoudi, 4–11
Decommodification of nourishment, 68, 144
Deforestation, 25
DeSantis, Ron, 40
Diabetes, 20, 59, 81
Donath, Jaap, 49
Doral, 53
Drought, 17, 25, 141
Dyer, Buddy, 118

E

Ecology, 4, 90, 121, 143
Ecosystem, 10, 91, 96, 142
  damage, 46
Education, 62, 83, 104
  about food, 68
Electronic Benefits Transfer (EBT) card, 149
Elevation, 48, 49, 54
El Paso, 121
Emergency
  Operations Center (EOC), 104
  planning, 15, 96
  response, 95, 139
Empowering Resilient Women, 101
Endometrial, 2, 20
Environmental Advisory Board, 98

F

Fairness, 7, 11
Farmers
  markets, 41, 58, 60–62, 65, 90, 99, 144, 148; increment in Orlando, 119; mobility of, 87, 137; network of, 100
  training, 122
Farming, 19, 25, 45, 85, 114, 119, 141–143
Federal Emergency Management Agency (FEMA), 49
Feeding America, 93, 99, 104, 148
Feeding South Florida, 56, 99, 130
Fertilizer, 142
Fishing, 64
Flood, 38, 46, 96
  control, 10
  insurance, 49, 91
  risk, 78, 87
Florida
  agriculture, 62–63; department of, 94
Food
  access, 1, 17–19, 19, 24, 113, 131; point, 62, 65, 68, 119, 131; transportation gaps, 118
  affordability, 17, 19, 29, 65
  availability, 17, 21, 25, 39, 145
  bank, 24, 27, 93–94, 99, 103, 119, 146, 148, 151, 153
  distribution, 64
  donation, 148, 151
  recovery, 66
  security, 17
  subsidization of, 79
  systems; threats, 62
  waste & recovery, 123
Food and Agricultural Organization (FAO), 17, 23
Food Environment Atlas, 55–56
Food Environment Briefs, 117
Food Policy Council, 60, 99, 119, 120, 151
  of Los Angeles, 120, 135
  of Madison, 123
  of Massachusetts, 113
  of Orlando, 119
Forbes.com, 49

Fortlauderdale.gov, 96
Fossil Fuel, 77, 98
Fresh Access Bucks Program (FAB), 99
Friedrich, Art, 61, 64, 99
Fruits, 2, 18, 25, 27, 59, 63, 90, 99, 149
Fünfgeld and McEvoy, 4, 8, 15

G
Gelber, Dan, 79
General Mills, 103
Gentrification, 44, 54, 86, 152
Gilbert, Jane, 80, 86–87
Gimenez, Carlos, 41, 81
Good Food Orlando, 119
Grain yield, 63
Great Depression, 43, 78
Greater Miami
 and the Beaches, 82
 Chamber of Commerce, 47
 mobility, 55
Greenhouse Gas (GHG), 7, 15, 25, 27, 66, 76, 94, 112
Green Infrastructure, 88–89
Green Living Advisory Board, 98
Green New Deal (GND), 78
Green Sustainable Building Ordinance, 84
Green Works Orlando, 118
Groundwater, 54, 91
Grow2heal, 66
Gunder, Valencia, 100

H
Habits, 43, 68, 83, 122
Haiti, 42, 43, 54, 102
Harvard Health Publishing, 2, 20
Health
 Foundation of South Florida, 62, 148
 implications, 59
 literacy, 67, 112, 149
 mental, 2, 20
Heifer's Locavore Index, 59
Heiman, Della, 61, 148
Hialeah, 43, 53, 84, 128, 146
Hidden Hungry, The, 56
Hightower, Marisa, 103
Hispanics, 43, 53
Holder, Cheryl, 50
Homelessness, 53
Horticulture therapy, 100
Horwitz, Jill, 94
Hot Pepper, 150
Housing, 53
Houston, Megan, 97
Human
 activity, 77
 impacts, 50, 76
Hunger Relief Plan, 97, 112
Hurricane, 14, 38, 43, 49, 101, 113, 136
 Andrew, 46
 Irma, 27, 68, 98, 100
 Katrina, 12, 27, 46, 104
Hydroponic systems, 142

I
Icma.org, 95
Initiative for A Competitive Inner City (ICIC), 3
Insightcced.org, 51
Institute for Food and Agricultural Standards (IFAS), 68, 85, 119, 142
Intergovernmental Panel on Climate Change (IPCC), 15, 49
International Institute for Sustainable Development (IISD), 23
International Research Institute (CGIAR), 25

International Water Resources
  Association (IWRA), 142–143

J
Jamaica, 42
Jesse Trice Community Health Center, 86
Jiler, James, 68, 100, 132
Jobs, 51, 55, 76, 112, 144
Johns Hopkins, 117–118, 134
Jurado, Jennifer, 90, 96

K
Kauffman.org, 48
Kidney, 2, 20
Knowledge, 11, 17, 68, 130, 150
Kresge Foundation, 102, 153

L
LaPradd, Charles, 45, 63–67, 86
Law, 93, 116, 139, 147
Legion Park, 99
Legislation, 105, 123, 139, 147
Levine, C. Daniella, 101
Little Haiti, 54
Livability, 37, 54, 76, 90, 95, 119
Logan International Airport, 113
Los Angeles, 120
  food; policy council, 121, 135; waste, 121, 150

M
Madison, 123
  Food Policy Council, 123
Maize, 25
Make the Homeless Smile, 100
Mango, 150

Marjory Stoneman Douglas High
  School, 93–94
Massachusetts Food Policy Council, 113
Massachusetts Local Food Action
  Plan, 113
Meal Gap report, 93
Methane, 27
Mexico, 63
Miami
  Beach, 53, 69, 82, 84–89
  City of, 86
  Climate Alliance (MCA), 100
  Dade County, 43–45, 48, 84, 86, 135–136
  Foundation, 101
  Gardens, 89
  International Airport, 48
  South of, 90
  Workers Center (MWC), 100
Miamidade.gov, 48
Miami-Dade, 37, 43–45, 48, 54, 80
  income range, 51
  urban planning, 62
Miamigov.com, 55
Microcephaly, 50
Milam's Market, 64
Milan Urban Food Policy Pact, 23, 86
Milanurbanfoodpolicypact.org, 23
Milk, 18, 25
Millet, 25
Mindbodygreen.Com, 104
Miramar, 96
Mitchell, Marc, 50
MIT U-Theory, 122
Mobility, 55, 83, 87, 137
Monroe, 80, 99
Mosquito, 81
Musculoskeletal disorder, 2, 20
Myboca.us, 98

N

NAFTA, 143
National Center for Biotechnology Information,, 19
National Climate Assessment (NCA), 14, 38
Natural disaster, 21, 27, 113, 134, 155
New York City Economic Development Corporation (NYCEDC), 116
Nicaragua, 42
Nieratka, R. Lindsey, 98
Non-Profit Organizations, 99
Nutrition Resilient Communities
design of, 111

O

Obama, Barack, 80
Obesity, 2, 19, 59, 89, 100, 150
Ocean acidification and warming, 46
Office of
Environmental Accountability and Transparency, 41
Food Initiative (OFI), 113, 128
Management and Budget, 37
Resilience and Sustainability, 84, 86, 88, 98
Opa-Locka, 44, 146
Orlando, 118–119
Overtown, 54

P

Palm Beach, 37, 41, 46, 56, 80, 83, 97, 135
Hunger Relief Plan, 112
Paris Climate Agreement, 77, 98
Parks, 89, 145
Patient Protection and Affordable Care Act, 67
Pembroke Pines, 96

People's Access to Community Horticulture (PATCH™), 92
Physical activity, 20, 89, 150
Political action, 77
Pompano Beach, 96
PowerU, 100
Presidente, 64
Public health, 54, 81, 136, 142
Publix, 64
Pumps, 88, 101
stopping floods, 9, 86

R

Recycling, 121
Red tide, 41
Regional Climate Action Plan (RCAP), 81, 83
Resalliance.org, 6
Resident Food Equity Advisors, 117
Resilience
action, 75
definition of, 4–6
Alliance, 6, 39
engineering, 4
of community, 12, 51
of economy, 76, 81, 86, 143
planning, 3, 23, 27, 114, 129, 136, 139
Resilient305.org, 38
Restaurant Food Waste Recycling Program, 121
Ridzuan, A.A., 13
Rockefeller Foundation, 8, 38, 40, 80, 82, 86, 102, 123, 134
Rolley, Otis, 151
Roosevelt, Franklin D., 78

S

Safe failure, 10
Saltwater intrusion, 63, 84, 95, 143
Sciencemarchmiami.org, 105

Scott, Rick, 40, 79
Sea-level rise (SLR), 47, 84, 112, 151
  adaptation, 86, 102
Snow birds, 42
Social
  activism, 145
  challenges, 51
  system, 4
Socio-Ecological Systems (SES), 4–6, 39
Sorghum, 25
South Africa, 122
Southeast Florida Regional Climate Change Compact, 40, 46, 80, 94, 102
South Florida
  Flood Control, 91
  food insecurity, 130
Soybeans, 25
Squire, Thi, 66–68, 104
Stoddard, Phillip, 80, 86, 90, 152–153
Suarez, Francis, 86
Sugar, 2, 19
Supplemental Nutrition Assistance Program (SNAP), 79
Sustainability
  Action Plan (SAP), 95
  Advisory Board, 95
  definition of, 6

T
Target, 100, 104
Tax base, 59, 86, 88, 90, 152
Tax Cuts and Jobs Act (2017), 137
Teamwork, 9, 152
Temperature, 21, 46, 48, 134, 142
Thepatchgarden.com, 93
Tinsley, Vanessa, 100
Torriente, Susanne, 82, 84–89, 101, 151
Tourism, 38, 47, 79, 143, 152
Trade associations, 139, 147

Trader Joe's, 64
Transformability, 4–8
Transportation, 25, 26, 53, 55, 95, 104–106, 112, 118, 139
  challenges, 76
  cost, 144
Trump, President, 40, 77
Tulane University, 18–20
Tyler and Moench, 6, 9–11, 15, 21

U
UN Food and Agriculture Organization, 27
United States
  Department of Agriculture (USDA), 65, 67
  Environmental Protection Agency, 27
Unitedwaymiami.org, 51, 56
U-Pick, 45, 65
Urban
  agriculture, 144; policy, 145
  Green Works (UGW), 100
  planning, 3, 7, 8, 16, 62, 153
  resilence, 8
Urbangreenworks.org, 100
Urbanization, 3, 39, 49, 82
Urbanland.uli.org, 88
USA Census, 41, 53
U-Theory, 122
Utilization, 17, 19–21

V
Vegetable, 2, 18, 27, 59, 63, 66, 90, 99, 101, 149, 154
  prescription, 100, 104, 131
Vildosola, D., 103
Vitamin deficiency, 2, 18
Von Dem Knesebeck, 2
Vulnerability assessment, 15, 84, 91, 134

W

Walmart, 64
Waste
  ammonia, 142
  recycling, 92
Wastewater, 54, 88
WeCount, 100
West Palm Beach, 97
Whitehouse.gov, 37
Whole Foods, 64
Wilson, DiVito, 68, 144
Wisconsin, 123
Withanachchi, S., 13, 131

Wnycstudios.org, 86
Wpb.org, 97
Wync, 86
Winn Dixie, 64
Wynwood Yard, 61

Z

Ziervogel and Ericksen, 21
Ziervogel and Frayne, 21
Zika virus, 50, 61
Zinc, 25

CPSIA information can be obtained
at www.ICGtesting.com
Printed in the USA
LVHW071348300122
709787LV00006B/242